PROBLEMS ON THE

DESIGN OF MACHINE ELEMENTS

PROBLEMS

on the

DESIGN OF MACHINE ELEMENTS

FOURTH EDITION

VIRGIL M. FAIRES
Professor of Mechanical Engineering
U. S. Naval Postgraduate School

ROY M. WINGREN
Professor of Mechanical Engineering
Texas A&M University

MACMILLAN PUBLISHING CO., INC.
New York

COLLIER MACMILLAN PUBLISHERS
London

MACMILLAN PUBLISHING CO., INC.
866 THIRD AVENUE, NEW YORK, NEW YORK 10022

COLLIER MACMILLAN CANADA, LTD.

PRINTED IN THE UNITED STATES OF AMERICA

Printing   16 17 18   Year   1 2 3 4

PREFACE TO FOURTH EDITION

NEARLY all of the problems in this new edition have been rewritten in order to conform to the latest design and engineering information. Even though it is often true that the data are the same as before, and even though the engineering answer is the same as, or virtually the same as, before, a major percentage of the *solutions* are different. While the applicable science has remained the same, some of the engineering has changed.

The appendix material has been expanded significantly, making the book even more convenient and helpful in the solution of its problems. Other teachers may find, as do the authors, that closed-textbook quizzes can be given, allowing the student use of the problem book, without imposing unduly on the student's memory. After a proper study of the textbook, a solution to most of the problems can be made by one with the "know-how" using only the information contained herein.

Noteworthy features, in addition to the expanded appendix, include: (1) a major increase in the emphasis on design; (2) a larger number of problems involving varying loading and fatigue; (3) a revision of problems to accord with new design information, notably for fits of parts, residual stresses, combines stresses (octahedral shear), shells, journal bearings, rolling bearings, bevel gearing, V-belts, roller chains, and welded joints; and (4) more attention to the heat treatment of steels. The problems tend to be comprehensive—that is, the attempt is often made to induce the student to think of several aspects of a particular situation. However, the questions are usually designated (a), (b), and so forth, allowing the teacher to assign only a part of a problem if he so desires. The authors made a conscious effort to avoid the "strength-of-materials type," wherein everything is given except the solution and the answer.

Naturally, the arrangement is such as to fit best with Faires' *Design of Machine Elements*, yet this work may be used in any course on machine design where the latest design approaches are to be featured. As before, there are unused numbers at the end of each section which the instructor may use for his own favorite problems.

The authors will be grateful to any who write of errors, omissions, or suggestions.

VMF
RMW

CONTENTS

SYMBOLS AND ABBREVIATIONS

Most of the symbols and abbreviations used in this book are listed below. In general these are the same as those used in the *Text*. See the *Text* for a complete list.

c_d	bearing clearance; c_d, diametral clearance; c_r, radial clearance
C	center distance; spring index; a constant
D	diameter; D_o, outside diameter; D_i, inside diameter
e	eccentricity of load; effective error in gear-tooth profiles; efficiency
h	height; a dimension; h_o, minimum film thickness in journal bearings
i	interference of metal
K_f	fatigue strength reduction factor
m_ω	velocity ratio
n	angular velocity; revolutions per minute; n_s, revolutions per second
N	design factor or factor of safety; sometimes, load normal to a surface
N_c	Number of coils; N_p, number of teeth in pinion; etc.
p	pressure in pounds per square inch
P	pitch; P_d, diametral pitch; P_c, circular pitch; maximum pressure
r	radius in inches
s	stress; s_u, ultimate strength; s_{us}, ultimate shear strength; etc.
S	Sommerfeld number
U	work; U_f, work of friction; U_s, work of spring
v	velocity; v_s, velocity in fps; v_m, velocity in fpm
δ (delta)	total elongation
θ (theta)	an angle
λ (lambda)	lead angle of worm or screw threads
μ (mu)	Poisson's ratio; absolute viscosity in reyns
π (pi)	$3.1416\ldots\ldots$
ρ (rho)	density
σ (sigma)	resultant normal stress in combined stresses; standard deviations
τ (tau)	resultant shearing stress in combined stresses
ω (omega)	angular velocity in radians per second
AISI	American Iron and Steel Institute
ALBA	American Leather Belting Association
ASTM	American Society of Testing Material
BHN	Brinell hardness number
CC	counterclockwise
CL	clockwise
fpm	feet per minute
fps	feet per second
hp	horsepower
ID	inside diameter
ksi	kips per square inch
OD	outside diameter
psi	pounds per square inch
rpm	revolutions per minute
SAE	Society of Automotive Engineers
OQT	oil quenched and tempered
QT	quenched and tempered
WQT	water quenched and tempered

PROBLEMS ON THE
DESIGN OF MACHINE ELEMENTS

Section 1

DESIGN FOR SIMPLE STRESSES

TENSION, COMPRESSION, SHEAR

DESIGN PROBLEMS

1. The link shown, made of AISI C1045 steel, as rolled, is subjected to a tensile load of 8000 lb. Let $h = 1.5b$. If the load is repeated but not reversed, determine the dimensions of the section with the design based on (a) ultimate strength, (b) yield strength. (c) If this link, which is 15 in. long, must not elongate more than 0.005 in., what should be the dimensions of the cross section?

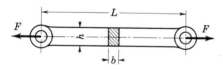

Problems 1-3.

2. The same as **1** except that the material is malleable iron, ASTM A47-52, grade 35 018.

3. The same as **1** except that the material is gray iron, ASTM 30.

4. A piston rod, made of AISI 3140 steel, OQT 1000°F (Fig. AF 2), is subjected to a repeated, reversed load. The rod is for a 20-in. air compressor, where the maximum pressure is 125 psig. Compute the diameter of the rod using a design factor based on (a) ultimate strength, (b) yield strength.

5. A hollow, short compression member, of normalized cast steel (ASTM A27-58, 65 ksi), is to support a load of 1500 kips with a factor of safety of 8 based on the ultimate strength. Determine the outside and inside diameters if $D_o = 2D_i$.

6. A short compression member with $D_o = 2D_i$ is to support a dead load of 25 tons. The material is to be 4130 steel, WQT 1100°F. Calculate the outside and inside diameters on the basis of (a) yield strength, (b) ultimate strength.

7. A round, steel tension member, 55 in. long, is subjected to a maximum load of 7000 lb. (a) What should be its diameter if the total elongation is not to exceed 0.030 in.? (b) Choose a steel that would be suitable on the basis of yield strength if the load is gradually applied and repeated (not reversed).

8. A centrifuge has a small bucket, weighing 0.332 lb. with contents, suspended on a manganese bronze pin (B138-A, ½ hard) at the end of a horizontal arm. If the pin is in double shear under the action of the centrifugal force, determine the diameter needed for 10,000 rpm of the arm. The center of gravity of the bucket is 12 in. from the axis of rotation.

CHECK PROBLEMS

9. The link shown is made of AISI C1020 annealed steel, with $b = \frac{3}{4}$ in. and $h = 1\frac{1}{2}$ in. (a) What force will cause breakage? (b) For a design factor of 4 based on the ultimate strength, what is the maximum allowable load? (c) If $N = 2.5$ based on the yield strength, what is the allowable load?

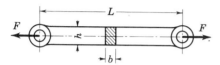

Problem 9.

10. A $\frac{3}{4}$-in. bolt, made of cold-finished B1113, has an effective stress area of 0.334 sq. in. and an effective grip length of 5 in. The bolt is to be loaded by tightening until the tensile stress is 80% of the yield strength, as determined by measuring the total elongation. What should be the total elongation?

11. A 4-lb. weight is attached by a $\frac{3}{8}$-in. bolt to a rotating arm 14 in. from the center of rotation. The axis of the bolt is normal to the plane in which the centrifugal force acts and the bolt is in double shear. At what speed will the bolt shear in two if it is made of AISI B1113, cold finish?

12. How many $\frac{3}{4}$-in. holes could be punched in one stroke in annealed steel plate of AISI C1040, $\frac{3}{16}$-in. thick, by a force of 60 tons?

13. What is the length of a bearing for a 4-in. shaft if the load on the bearing is 6400 lb. and the allowable bearing pressure is 200 psi of the projected area?

BENDING STRESSES

DESIGN PROBLEMS

14. A lever keyed to a shaft is $L = 15$ in. long and has a rectangular cross section of $h = 3t$. A 2000-lb. load is gradually applied and reversed at the end as shown; the material is AISI C1020, as rolled. Design for both ultimate and

yield strengths. (a) What should be the dimensions of a section at $a = 13$ in.? (b) at $b = 4$ in.? (c) What should be the size where the load is applied?

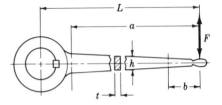

Problem 14.

15. A simple beam 54 in. long with a load of 4 kips at the center is made of cast steel, SAE 080. The cross section is rectangular (let $h \approx 3b$). (a) Determine the dimensions for $N = 3$ based on the yield strength. (b) Compute the maximum deflection for these dimensions. (c) What size may be used if the maximum deflection is not to exceed 0.03 in.?

16. The same as **15**, except that the beam is to have a circular cross section.

17. A simple beam, 48 in. long, with a static load of 6000 lb. at the center, is made of C1020 structural steel. (a) Basing your calculations on the ultimate strength, determine the dimensions of the rectangular cross section for $h = 2b$. (b) Determine the dimensions based on yield strength. (c) Determine the dimensions using the principle of "limit design."

18. The bar shown is subjected to two vertical loads, F_1 and F_2, of 3000 lb. each, that are $L = 10$ in. apart and 3 in. (a,d) from the ends of the bar. The design factor is 4 based on the ultimate strength; $h = 3b$. Determine the dimensions h and b if the bar is made of (a) gray cast iron, SAE 111; (b) malleable cast iron, ASTM A47-52, grade 35 018; (c) AISI C1040, as rolled (Fig. AF 1).

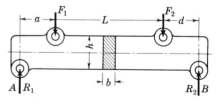

Problems 18, 19.

Sketch the shear and moment diagrams approximately to scale.

19. The same as **18**, except that F_1 acts up (F_2 acts down).

20. The bar shown, supported at A and B, is subjected to a static load F of 2500 lb. at $\theta = 0$. Let $d = 3$ in., $L = 10$ in. and $h = 3b$. Determine the dimensions of the section if the bar is made of (a) gray iron, SAE 110; (b) malleable cast iron, ASTM A47-52, grade 32 510; (c) AISI C1035 steel, as rolled. (d) For economic reasons, the pins at A, B, and C are to be the same size. What should be their diameter if the material is AISI C1035, as rolled, and the mounting is such that each is in double shear? Use the basic dimensions from (c) as needed. (e) What sectional dimensions would be used for the C1035 steel if the principle of "limit design" governs in (c)?

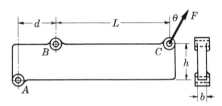

Problems 20, 21.

21. The same as **20**, except that $\theta = 30°$. Pin B takes all the horizontal thrust.

22. A cast-iron beam, ASTM 50, as shown, is 30 in. long and supports two gradually applied, repeated loads (in phase), one of 2000 lb. at $e = 10$ in. from the free end, and one of 1000 lb. at the free end. (a) Determine the dimensions of the cross section if $b = c \approx 3a$. (b) The same as (a) except that the top of the tee is below.

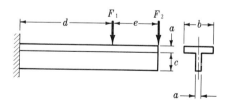

Problem 22.

CHECK PROBLEMS

23. An I-beam is made of structural steel, AISI C1020, as rolled. It has a depth of 3 in. and is subjected to two loads: F_1 and $F_2 = 2F_1$; F_1 is 5 in. from one end and F_2 is 5 in. from the other end, with the supports on the outside ends. The beam is 25 in. long; flange width is $b = 2.509$ in.; $I_x = 2.9$ in.[4] Determine (a) the approximate values of the load to cause elastic failure, (b) the safe loads for a factor of safety of 3 based on the yield strength, (c) the safe load allowing for flange buckling (§ 1.24), (f) the maximum deflection caused by the safe loads.

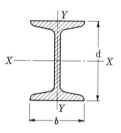

Problems 23-25.

24. The same as **23**, except that the material is aluminum alloy, 2024-T4, heat treated.

25. A light I-beam is 80 in. long, simply supported, and carries a static load at the midpoint. The cross section has a depth of $d = 4$ in., a flange width of $b = 2.66$ in., and $I_x = 6.0$ in.[4] (see figure). (a) What load will the beam support if it is made of C1020, as-rolled steel, and flange buckling (§ 1.24) is considered? (b) Consider the stress owing to the weight of the beam, which is 7.7 lb./ft., and decide whether or not the safe load should be less.

26. What is the stress in a band-saw blade due to being bent around a $13\frac{3}{4}$-in. pulley? The blade thickness is 0.0265 in. (Additional stresses arise from the initial tension and forces of sawing.)

27. A cantilever beam of rectangular cross section is tapered so that the depth varies uniformly from 4 in. at the fixed end to 1 in. at the free end. The width is 2 in. and the length 30 in. What safe load, acting repeatedly with minor shock, may be applied to the free end? The material is AISI C1020, as rolled.

TORSIONAL STRESSES

DESIGN PROBLEMS

28. A centrifugal pump is to be driven by a 15-hp electric motor at 1750 rpm. What should be the diameter of the pump shaft if it is made of AISI C1045 as rolled? Consider the load as gradually repeated.

29. A shaft in torsion only is to transmit 2500 hp at 570 rpm with medium shock. Its material is AISI 1137 steel, annealed. (a) What should be the diameter of a solid shaft? (b) If the shaft is hollow, $D_o = 2D_i$, what size is required? (c) What is the weight per foot of length of each of these shafts? Which is the lighter? By what percentage? (d) Which shaft is the more rigid? Compute the torsional deflection of each for a length of 10 ft.

30. The same as **29,** except that the material is AISI 4340, OQT 1200°F.

31. A steel shaft is transmitting 40 hp at 500 rpm with minor shock. (a) What should be its diameter if the deflection is not to exceed 1° in $20D$? (b) If deflection is primary what kind of steel would be satisfactory?

32. A square shaft of cold-finish AISI 1118 transmits a torsional moment of 1200 in-lb. For medium shock, what should be its size?

CHECK PROBLEMS

33. A punch press is designed to exert a force sufficient to shear a $1\frac{5}{16}$-in. hole in a $\frac{1}{2}$-in. steel plate, AISI C1020, as rolled. This force is exerted on the shaft at a radius of $\frac{3}{4}$-in. (a) Compute the torsional stress in the 3.5-in. shaft (bending neglected). (b) What will be the corresponding design factor if the shaft is made of cold-rolled AISI 1035 steel (Table AT 10)? Considering the shock loading that is characteristic of this machine, do you think the design is safe enough?

34. The same as **33,** except that the shaft diameter is $2\frac{3}{4}$ in.

35. A hollow annealed Monel propeller shaft has an external diameter of $13\frac{1}{2}$ in. and an internal diameter of $6\frac{1}{2}$ in.; it transmits 10,000 hp at 200 rpm. (a) Compute the torsional stress in the shaft (stress from bending and propeller thrust are not considered). (b) Compute the factor of safety. Does it look risky?

STRESS ANALYSIS

DESIGN PROBLEMS

36. A hook is attached to a plate as shown and supports a static load of 12,000 lb. The material is to be AISI C1020, as rolled. (a) Set up strength equations for dimensions d, D, h, and t. Assume that the bending in the plate is negligible. (b) Determine the minimum permissible value of these dimensions. In estimating the strength of the nut, let $D_1 = 1.2d$. (c) Choose standard fractional dimensions which you think would be satisfactory.

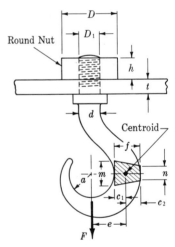

Problems 36-38.

37. The same as **36,** except that a shock load of 4000 lb. is repeatedly applied.

38. The connection between the plate and hook, as shown, is to support a load F. Determine the value of dimensions D, h, and t in terms of d if the connection is to be as strong as the rod of diameter d. Assume that $D_1 = 1.2d$,

$s_{us} = 0.75s_u$, and that bending in the plate is negligible.

39. (a) For the connection shown, set up strength equations representing the various methods by which it might fail. Neglect bending effects. (b) Design this connection for a load of 2500 lb. Both plates and rivets are of AISI C1020, as rolled. The load is repeated and reversed with mild shock. Make the connection equally strong on the basis of yield strengths in tension, shear, and compression.

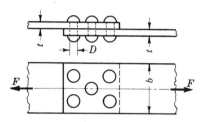

Problems 39, 40.

40. The same as **39,** except that the material is 2024–T4, aluminum alloy.

41. (a) For the connection shown, set up strength equations representing the various methods by which it might fail. (b) Design this connection for a load of 8000 lb. Use AISI C1015, as rolled, for the rivets, and AISI C1020, as rolled, for the plates. Let the load be repeatedly applied with minor shock in one direction and make the connection equally strong on the basis of ultimate strengths in tension, shear, and compression.

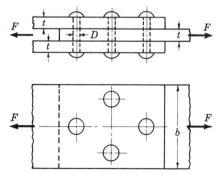

Problem 41.

42. Give the strength equations for the connection shown, including that for the shear of the plate by the cotter.

43. A steel rod, as-rolled AISI C1035, is fastened to a $\frac{7}{8}$-in., as-rolled C1020 plate by means of a cotter that is made of as-rolled C1020, in the manner shown. (a) Determine all dimensions of this joint if it is to withstand a reversed shock load $F = 10$ kips, basing the design on yield strengths. (b) If all fits are free-running fits, decide upon tolerances and allowances.

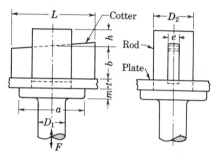

Problems 42-44.

44. A 1-in. (D_1) steel rod (as-rolled AISI C1035) is to be anchored to a 1-in. steel plate (as-rolled C1020) by means of a cotter (as-rolled C1035) as shown. (a) Determine all the dimensions for this connection so that all parts have the same ultimate strength as the rod. The load F reverses direction. (b) Decide upon tolerances and allowances for loose-running fits.

45. (a) Give all the simple strength equations for the connection shown. (b) Determine the ratio of the dimensions a, b, c, d, m, and n to the dimension D so that the connection will be equally strong in tension, shear, and compression. Base the calculations on ultimate strengths and assume $s_{us} = 0.75s_u$.

46. The same as **45,** except that the calculations are to be based on yield strengths. Let $s_{sy} = 0.6s_y$.

47. Design a connection similar to the one shown for a gradually applied and reversed load of 12 kips. Base design stresses on yield strengths and let the material be AISI C1040 steel, annealed.

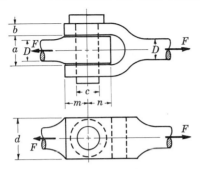

Problems 45-47.

Examine the computed dimensions for proportion, making changes that you deem advisable.

48. Give all the strength equations for the union of rods shown.

49–68. Design a union-of-rods joint similar to that shown for a reversing load and material given in the accompanying table. The taper of the cotter is to be ½ in. in 12 in. (see **172**). (a) Using design stresses based on yield strengths, determine all dimensions to satisfy the necessary strength equations. (b) Modify dimensions as necessary for good proportions, being careful not to weaken the joint. (c) Decide upon tolerances and allowances for loose fits. (d) Sketch to scale each part of the joint showing all dimensions needed for manufacture, with tolerance and allowances.

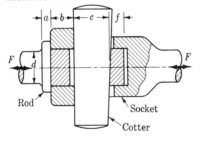

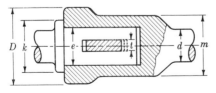

Problems 48-68.

CHECK PROBLEMS

69. The connection shown has the following dimensions: $d = 1\frac{1}{4}$ in., $D = 2\frac{1}{2}$ in., $D_1 = 1\frac{1}{2}$ in., $h = \frac{5}{8}$ in., $t = \frac{1}{2}$

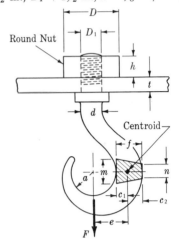

Problem 69.

Prob. No.	Load, lb.	AISI No. As Rolled
49	3000	1020
50	3500	1030
51	4000	1117
52	4500	1020
53	5000	1015
54	5500	1035
55	6000	1040
56	6500	1020
57	7000	1015
58	7500	1118
59	8000	1022
60	8500	1035
61	9000	1040
62	9500	1117
63	10,000	1035
64	10,500	1022
65	11,000	1137
66	11,500	1035
67	12,000	1045
68	12,500	1030

in.; it supports a load of 15 kips. Compute the tensile, compressive, and shear stresses induced in the connection. What is the corresponding design factor based on the yield strength if the rod and nut are made of AISI C1045, as rolled, and the plate is structural steel (1020)?

70. In the figure, let $D = \frac{3}{4}$ in., $t = \frac{7}{16}$ in., $b = 3\frac{3}{4}$ in., and let the load, which is applied centrally so that it tends to pull the plates apart, be 15 kips. (a) Compute the stresses in the various parts of the connection. (b) If the material is AISI C1020, as rolled, what is the design factor of the connection based on yield strengths?

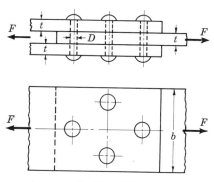

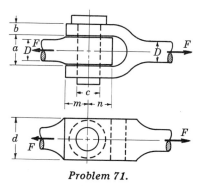

Problem 70.

71. For the connection shown, let $a = 1\frac{5}{16}$ in., $b = \frac{9}{16}$ in., $c = \frac{3}{4}$ in., $d = 1\frac{1}{2}$ in., $D = \frac{3}{4}$ in., $m = n = \frac{15}{16}$ in. The material is AISI C1040, annealed (see Fig. AF 1). (a) For a load of 7500 lb., compute the various tensile, compressive, and shear stresses. Determine the factor of safety based on (b) ultimate strengths, (c) yield strengths.

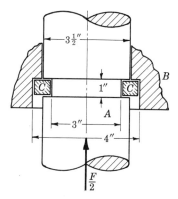

Problem 71.

72. The upper head of a 60,000-lb. tensile-testing machine is supported by two steel rods, one of which A is shown. These rods A are attached to the head B by split rings C. The test specimen is attached to the upper head B so that the tensile force in the specimen pulls down on the head and exerts a compressive force on the rods A. When the machine is exerting the full load, compute (a) the compressive stress in the rods, (b) the bearing stress between the rods and the rings, (c) the shearing stress in the rings.

Problem 72.

DEFORMATIONS

73. A load of 22,000 lb. is gradually applied to a 2-in. round rod, 10 ft. long. The total elongation is observed to be 0.03 in. If the stretching is entirely elastic, (a) what is the modulus of elasticity, and (b) what material would you judge it to be, wrought iron or stainless steel (from information available in the tables)? (c) How much energy is absorbed by the rod? (d) Suppose that the material is aluminum alloy 3003-H14; compute its total elongation for the same load. Is this within elastic action?

74. The same as **73**, except that $F = 88$ kips and total $\delta = 0.112$ in. Is the computation for part (d) valid? Explain.

75. (a) A square bar of SAE 1020, as rolled, is to carry a tensile load of 40 kips. The bar is to be 4 ft. long. A design factor of 5 based on the ultimate stress is desired. Moreover, the total deformation should not exceed 0.024 in.

What should be the dimensions of the section? (b) Using SAE 1045, as rolled, but with the other data the same, find the dimensions. (c) Using SAE 4640, OQT 1000°F, but with other data the same as in (a), find the dimensions. Is there a change in dimensions as compared with part (b)? Explain the difference or the lack of difference in the answers.

76. The steel rails on a railroad track are laid when the temperature is 40°F. The rails are welded together and held in place by the ties so that no expansion is possible due to temperature changes. What will be the stress in the rails when heated by the sun to 120°F (§ 1.29)?

77. Two steel rivets are inserted in a riveted connection. One rivet connects plates that have a total thickness of 2 in., while the other connects plates with a total thickness of 3 in. If it is assumed that, after heading, the rivets cool from 600°F and that the coefficient of expansion as given in the *Text* applies, compute the stresses in each rivet after it has cooled to a temperature of 70°F, (no external load). See § 1.29. Also assume that the plates are not deformed under load. Is such a stress likely? Why is the actual stress smaller?

78. Three flat plates are assembled as shown; the center one *B* of chromium steel, AISI 5140 OQT 1000°F, and the outer two *A* and *C* of aluminum alloy 3003-H14, are fastened together so that they will stretch equal amounts. The steel plate is $2 \times \frac{1}{2}$ in., the aluminum plates are each $2 \times \frac{1}{8}$ in., $L = 30$ in., and the load is 24,000 lb. Determine (a) the stress in each plate, (b) the total elongation, (c) the energy absorbed by the steel plate if the load is gradually applied, (d) the energy absorbed by the aluminum plate. (e) What will be the

stress in each plate if in addition to the load of 24,000 lb. the temperature of the assembly is increased by 100°F?

79. The same as **78**, except that the outer plates are aluminum bronze, B150-1, annealed.

80. A machine part as shown is made of AISI C1040, annealed steel; $L_1 = 15$ in., $L_2 = 6$ in., $D_1 = \frac{3}{4}$ in., and $D_2 = \frac{1}{2}$ in. Determine (a) the elongation due to a force $F = 6000$ lb., (b) the energy absorbed by each section of the part if the load is gradually applied.

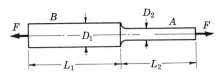

Problems 80, 81.

81. A rod as shown is made of AISI 2340 steel, OQT 1000°F, and has the following dimensions: $L_1 = 20$ in., $L_2 = 12$ in., $D_1 = \frac{7}{8}$ in., and $D_2 = \frac{3}{4}$ in. The unit strain at point A is measured with a strain gage and found to be 0.0025 in./in. Determine (a) the total deformation, and (b) the force on the rod.

82. A rigid bar H is supported as shown in a horizontal position by the two rods (aluminum 2024 T4, and steel AISI 1045, as rolled), whose ends were both in contact with H before loading was applied. The ground and block B are also to be considered rigid. What must be the cross-sectional area of the steel rod if, for the assembly, $N = 2$ based on the yield strengths?

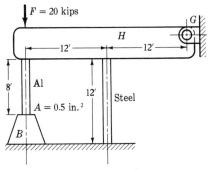

Problem 82.

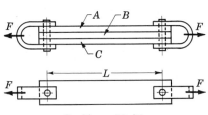

Problems 78, 79.

83. The bar shown supports a static load $F = 2.5$ kips with $\theta = 0$; $d = 3$ in., $L = 10$ in., $h = 2\frac{3}{4}$ in., $b = 1$ in. It is made of AISI 1035, as rolled. (a) How far does point C move upon gradual application of the load if the movement of A and B is negligible? (b) How much energy is absorbed?

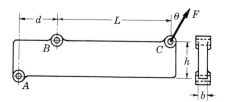

Problem 83.

PRESSURE VESSELS

84. A storage tank for air, 36 in. in diameter, is to withstand an internal pressure of 200 psi with a design factor of 4 based on s_u. The steel has the strength equivalent of C1020 annealed and the welded joints should have a relative strength (efficiency) of 90%. Determine a suitable plate thickness. Compute the stress on a diametral section and compare it with the longitudinal stress.

85. A spherical air tank stores air at 3000 psig. The tank is to have an inside diameter of 7 in. (a) What should be the wall thickness and weight of the tank if it is made of 301, $\frac{1}{4}$-hard, stainless steel, with a design factor of 1.5 based on the yield strength and a joint efficiency of 90%? (b) Compute the wall thickness and weight if annealed titanium (B265, gr. 5) is used? (c) What is the additional saving in weight if the titanium is hardened? Can you think of circumstances for which the higher cost of titanium would be justified?

86. Decide upon a material and estimate a safe wall thickness of a cylindrical vessel to contain helium at $-300°F$ and 2750 psi. The welded joint should have a relative strength $\leqq 87\%$, and the initial computations are to be for a 12-in.-diameter, 30-ft.-long tank. (*Note:* Mechanical properties of metals at this low temperature are not available in the *Text*. Refer to *INCO Nickel Topics*, vol. 16, no. 7, 1963, or elsewhere.)

CONTACT STRESSES

87. (a) A 0.75-in. diameter roller is in contact with a plate-cam surface whose width is 0.5 in. The maximum load is 2.5 kips where the radius of curvature of the cam surface is 3.333 in. Compute the Hertz compressive stress. (b) The same as (a) except that the follower has a plane flat face. (c) The same as (a) except that the roller runs in a grooved face and contacts the concave surface. (d) What is the maximum shear stress for part (a) and how far below the surface does it exist?

88. Two 20° involute teeth are in contact along a "line" where the radii of curvature of the profiles are respectively 1.03 and 3.42 in. The face width of the gears is 3 in. If the maximum permissible contact stress for carburized teeth is 200 ksi, what normal load may these teeth support?

TOLERANCES AND ALLOWANCES

89. The pin for a yoke connection has a diameter D of $\frac{3}{4}$ in., a total length of $2\frac{1}{2}$ in., with a head that is $1\frac{1}{4}$ in. in diameter and $\frac{3}{8}$ in. thick. The tolerance on D (both pin and hole) is 0.003 in., with an allowance of 0.001 in., basic-hole system. Sketch the pin showing all dimensions with appropriate tolerances.

90. A shaft with a nominal diameter of 8 in. is to fit in a hole. Specify the allowance, tolerances, and the limit diameters of the shaft and hole on a sketch for: (a) a close sliding fit, (b) a precision-running fit, (c) medium-running fit, (d) a loose-running fit.

91. The same as **90**, except that the nominal diameter is 4 in.

92. A cast-iron gear is to be shrunk onto a 3-in. steel shaft. (a) Determine the tolerance and the maximum, minimum, and average interferences of metal for class FN 1 fit. (b) Sketch and dimension the shaft and hole with proper tolerances. (c) Compute the stresses by the method given in the *Text* (§ 3.8) for the maximum and minimum interferences of metal.

93. The same as **92**, except that the gear hub is C1035 steel and class of fit is FN 3.

94. For a No. 7 ball bearing, the *New Departure Handbook* states that the maximum bore should be 1.3780 in. and the minimum, 1.3775 in.; for average conditions, the shaft should have a maximum diameter of 1.3784 in. and a minimum of 1.3779 in. (a) Determine the corresponding tolerances and allowances. (b) What class of fit is this? (c) New Departure states: ". . . bearing bores are held uniformly close, . . . averaging within 1.3778 in. to 1.3776 in." What will be the maximum and minimum interference of metal with these diameters (if maximum and minimum sizes are deliberately chosen for assembly)?

95. For a roller bearing having a bore of 65 mm., an SKF catalog states that the largest diameter should be 2.5591 in. and the smallest, 2.5585 in. If this bearing is to be used in a gear transmission, it is recommended for the shaft (where the bearing fits) to have a maximum diameter of 2.5600 in. and a minimum of 2.5595 in. (a) Determine the tolerances and allowances (or interferences of metal) for this installation. (b) What class of fit would this be?

TOLERANCES, STATISTICAL CONSIDERATIONS

96. (a) A machine tool is capable of machining parts so that the standard deviation of one critical dimension is 0.0006 in. What minimum tolerance may be specified for this dimension if it is expected that practically all of the production be acceptable? Assume that it is possible to "center the process." (b) The same as (a), except that it has been decided to tolerate approximately 4.56% scrap.

97. A pin and the hole into which it fits have a nominal diameter of 1½ in. The pin tolerance has been set at 0.002 in., the bore tolerance at 0.003 in., and the allowance at 0.001 in., basic hole system. The parts are to be made on machines that give a natural spread of 0.0015 in. for the pin and 0.002 in. for the hole. Assuming that the processes are centered, determine the expected minimum clearance and the maximum clearance. What is the most frequent clearance?

98. A rod and the hole into which it fits has a nominal diameter of 2 in. The tolerances are 0.003 in. for both rod and hole, and the allowance as 0.001 in., basic hole system. The natural spread of the process of manufacturing the hole is 0.002 in., and for the rod, 0.0015 in. What are the probable maximum and minimum clearances, provided that the tolerances are met, but assuming that the processes might simultaneously operate at their extreme permissible position?

99. It is desired that the clearance in a 4-in. bearing neither exceed 0.004 in. nor be less than 0.002 in. Assume that the natural spread of the processes by which the journal and the bearing surfaces are finished is the same. (a) What should be the natural spread of these processes? (b) Assuming this natural spread to be equal to the tolerance, determine the corresponding allowance. (c) If the foregoing conditions are not practical decide upon practical tolerances and allowances for the computed natural spread.

100. A 4-in. journal-bearing assembly is made for a class RC 6 fit. Assume that the natural spread of the manufacturing process will be about 75% of the tolerance. Compute the probable maximum and minimum clearances (which occur when the processes are not centered) and compare with the allowance. Make a sketch of the journal and hole properly dimensioned.

101. The same as **100**, except that class RC 3 fit is used.

102. It is desired that the running clearance for a 3-in. bearing be between approximately 0.003 in. and 0.007 in. The natural spread of the processes of finishing the journal and bearing are expected to be virtually the same ($\sigma_1 = \sigma_2$). Decide upon a suitable tolerance and allowance with a sketch properly dimensioned (to a ten thousandth). (Suggestion: compute first a theoretical natural spread for bearing and journal from the given spread of the clearances. Let the tolerances be approximately equal to this computed NS, and assume that manufacturing processes are available that produce an actual NS of about 70% of this computed NS.) Check for processes being off center but within $\pm 3\sigma$ limits so that virtually no scrap is manufactured.

103. If the tolerances shown are maintained during manufacture, say with the processes centered, what would be the approximate overall tolerance and limit dimensions?

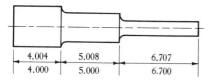

| 4.004 | 5.008 | 6.707 |
| 4.000 | 5.000 | 6.700 |

Problem 103.

104. If a cylindrical part needs to have the following tolerances, what process would you recommend for finishing the surface in each instance? (a) 0.05 in., (b) 0.01 in., (c) 0.005 in., (d) 0.001 in., (e) 0.0001., (f) 0.00005 in.?

105. If it cost $100 to finish a certain surface to 500 microinches rms, what would be the approximate cost to finish it to the following roughness: (a) 125, (b) 32, (c) 8, (d) 2 μin. rms?

DATA LACKING—DESIGNER'S DECISIONS*

106–125. Design a bell crank, similar to the one shown, to carry a mild shock load. The mechanical advantage $(L_1/L_2 = F_2/F_1)$, the force F_1, the length L_1, and the material are given in the

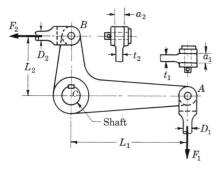

Problems 106-125.

* Properties of rolled structural sections are found in various handbooks.

accompanying table. (a) Make all significant decisions, including tolerances and allowances. One approach could be to compute dimensions of the yoke connections first; t should be a little less than a. An assumption for the shaft may be that, on occasion, the torque for F_1 is transmitted through the shaft (ignoring bending for local convenience). (b) Check all dimensions for good proportion; modify as desirable. (c) Sketch to scale each part, showing all dimensions with tolerances necessary to manufacture.

Prob. No.	Load F_1	L_1	AISI No. As Rolled	Mech. Advantage
106	700	12	C1020	1.5
107	650	14	C1020	2
108	600	15	C1022	2.5
109	550	18	C1035	3
110	500	20	C1040	4
111	800	12	C1020	1.5
112	750	14	C1020	2
113	700	15	C1022	2.5
114	650	18	C1035	3
115	600	20	C1040	4
116	900	12	C1020	1.5
117	850	14	C1020	2
118	800	15	C1022	2.5
119	750	18	C1035	3
120	700	20	C1040	4
121	1000	12	C1020	1.5
122	950	14	C1020	2
123	900	15	C1022	2.5
124	850	18	C1035	3
125	800	20	C1040	4

126. A simple beam 12 ft. long is to support a concentrated load of 10 kips at the midpoint with a design factor of at least 2.5 based on yield strength. (a) What is the size and weight of the lightest steel (AISI C1020, as rolled) I-beam that can be used? (b) Compute its maximum deflection. (c) What size beam should be used if the deflection is not to exceed $\frac{1}{4}$ in.?

127. A 10-in., 35-lb. I-beam is used as a simple beam, supported on 18-ft. centers, and carrying a total uniformly distributed load of 6000 lb. Determine the maximum stress and the maximum deflection.

128. The same as **127**, except the beam is a cantilever.

129. Two equal angles, placed back to back as shown, act as a simple beam and are to support a load of $F = 2{,}000$ lb.; $L = 40$ in.; $a = 15$ in. What size angles should be used if the maximum stress is not to exceed 20 ksi? The stress due to the weight of the angles is negligible.

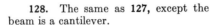
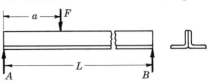

Problems 129, 130.

130. The same as **129**, except that a rolled T-section is to be used.

131–140. Unused numbers are left at the end of each section; the instructor may use them for his favorite problems or for variations of problems in this book.

Section 2

VARYING LOADS AND STRESS CONCENTRATIONS

VARYING STRESSES—NO CONCENTRATION

DESIGN PROBLEMS

141. The maximum pressure of air in a 20-in. cylinder (double-acting air compressor) is 125 psig. What should be the diameter of the piston rod if it is made of AISI 3140, OQT at 1000°F, and if there are no stress raisers and no column action? Let $N = 1.75$; indefinite life desired. How does your answer compare with that obtained for **4**?

142. A link as shown is to be made of AISI 2330, WQT 1000°F. The load $F = 5$ kips is repeated and reversed. For the time being, ignore stress concentrations. (a) If its surface is machined, what should be its diameter for $N = 1.40$. (b) The same as (a), except that the surface is mirror polished. What would be the percentage saving in weight? (c) The same as (a), except that the surface is as-forged.

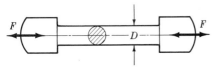

Problems 142-144.

143. The same as **142,** except that, because of a corrosive environment, the link is made from cold-drawn silicon bronze B and the number of reversals of the load is expected to be less than 3×10^7.

144. The same as **142,** except that the link is made of aluminum alloy 2024-T4 with a minimum life of 10^7 cycles.

145. A shaft supported as a simple beam, 18 in. long, is made of carburized AISI 3120 steel (Table AT 10). With the shaft rotating, a steady load of 2000 lb. is applied midway between the bearings. The surfaces are ground. Indefinite life is desired with $N = 1.6$ based on endurance strength. What should be its diameter if there are no surface discontinuities?

146. (a) A lever as shown with a rectangular section is to be designed for indefinite life and a reversed load of $F = 900$ lb. Find the dimensions of a section without discontinuity where $b = 2.8t$ and $L = 14$ in. for a design factor of $N = 2$. The material is AISI C1020, as rolled, with an as-forged surface. (b)

13

Compute the dimensions at a section where $e = 4$ in.

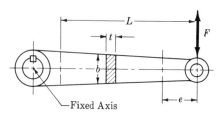

Problems 146, 147.

147. The same as **146**, except that the reversals of the load are not expected to exceed 10^5 (Table AT 10).

148. A shaft is to be subjected to a maximum reversed torque of 15,000 in-lb. It is machined from AISI 3140 steel, OQT 1000°F (Fig. AF 2). What should be its diameter for $N = 1.75$?

149. The same as **148**, except that the shaft is hollow with the outside diameter twice the inside diameter.

150. The link shown is machined from AISI 1035 steel, as rolled, and subjected to a repeated tensile load that varies from zero to 10 kips; $h = 1.5b$. (a) Determine these dimensions for $N = 1.40$ (Soderberg) at a section without stress concentration. (b) How much would these dimensions be decreased if the surfaces of the link were mirror polished?

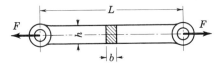

Problems 150, 151, 158.

151. The same as **150**, except that the link operates in a brine solution. (*Note:* The corroding effect of the solution takes precedence over the surface finish.)

152. The simple beam shown, 30-in. long $(=a + L + d)$, is made of AISI C1022 steel, as rolled, left as forged. At $a = 10$ in., $F_1 = 3000$ lb. is a dead load. At $d = 10$ in., $F_2 = 2400$ lb. is a repeated, reversed load. For $N = 1.5$,

indefinite life, and $h = 3b$, determine b and h. (Ignore stress concentration).

153. The same as **152**, except that the cycles of F_2 will not exceed 100,000 and all surfaces are machined.

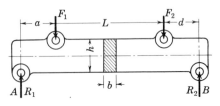

Problems 152, 153.

154. A round shaft, made of cold-finished AISI 1020 steel, is subjected to a variable torque whose maximum value is 6283 in-lb. For $N = 1.5$ on the Soderberg criterion, determine the diameter if (a) the torque is reversed, (b) the torque varies from zero to a maximum, (c) the torque varies from 3141 in-lb. to a maximum.

CHECK PROBLEMS

155. A simple beam 2 ft. long is made of AISI C1045 steel, as rolled. The dimensions of the beam, which is set on edge, are 1 in. $\times$ 3 in. At the midpoint is a repeated, reversed load of 4000 lb. What is the factor of safety?

156. The same as **155**, except that the material is normalized and tempered cast steel, SAE 080.

157. A $1\frac{1}{2}$-in. shaft is made of AISI 1045 steel, as rolled. For $N = 2$, what repeated and reversed torque can the shaft sustain indefinitely?

VARIABLE STRESSES WITH STRESS CONCENTRATIONS

DESIGN PROBLEMS

NOTE. *In determining the actual fatigue factor K_f, estimate it as well as possible from data at hand.*

158. The load on the link shown (**150**) is a maximum of 10 kips, repeated and reversed. The link is forged from AISI C1020, as rolled, and it has a $\frac{1}{4}$-in. drilled hole on the center line of the wide side. Let $h = 2b$ and $N = 1.5$. Determine b and h at the hole (no column action) (a) for indefinite life, (b) for 50,000 repetitions (no reversal) of the maximum load, (c) for indefinite life but with a ground and polished surface. In this case, compute the maximum stress.

159. A connecting link is as shown, except that there is a $\frac{1}{8}$-in. radial hole drilled through it at the center section. It is machined from AISI 2330, WQT 1000°F, and it is subjected to a repeated, reversed axial load whose maximum value is 5 kips. For $N = 1.5$, determine the diameter of the link at the hole (a) for indefinite life; (b) for a life of 10^5 repetitions (no column action). (c) In the link found in (a) what is the maximum tensile stress?

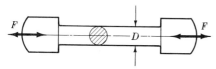

Problem 159.

160. A machine part of uniform thickness $t = b/2.5$ is shaped as shown and machined all over from AISI C1020, as rolled. The design is for indefinite life for a load repeated from 1750 lb. to 3500 lb. Let $d = b$. (a) For a design factor of 1.8 (Soderberg), what should be the dimensions of the part? (b) What is the maximum tensile stress in the part as designed?

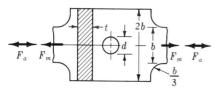

Problems 160, 161.

161. The same as **160**, except that the cycles of loading should not exceed 10^5.

162. The beam shown has a circular cross section and supports a load F that varies from 1000 lb. to 3000 lb.; it is machined from AISI C1020 steel, as rolled. Determine the diameter D if $r = 0.2D$ and $N = 2$; indefinite life.

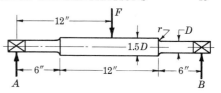

Problems 162-164.

163. The same as **162**, except that the load F is steady at $F = 3000$ lb. and the beam rotates as a shaft.

164. The shaft shown is machined from C1040, OQT 1000°F (Fig. AF 1). It is subjected to a torque that varies from zero to 10,000 in-lb. $(F = 0)$. Let $r = 0.2D$ and $N = 2$. Compute D. What is the maximum torsional stress in the shaft?

165. An axle (nonrotating) is to be machined from AISI 1144, OQT 1000°F, to the proportions shown, with a fillet radius $r \approx 0.25D$; F varies from 400 lb. to 1200 lb.; the supports are to the left of BB, not shown. Let $N = 2$ (Soderberg line). (a) At the fillet, compute D and the maximum tensile stress. (b) Compute D at section BB. (c) Specify suitable dimensions. Keeping the given proportions, would a smaller diameter be permissible if the fillet were shot-peened?

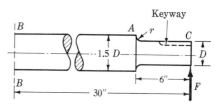

Problems 165-167.

166. A pure torque varying from 5 in-kips to 15 in-kips is applied at section C $(F = 0)$ of the machined shaft shown. The fillet radius $r = D/8$ and the torque passes through the profile keyway at C. The material is AISI 1050, OQT 1100°F, and $N = 1.6$. (a) What should be the diameter? (b) If the fillet radius were increased to $D/4$ would it be reasonable to use a smaller D?

167. The same as **166**, except that the torque varies from -5 in-kips to $+15$ in-kips.

168. A cantilever beam as shown is to be subjected to a reversing load of 3000 lb. Let the radius of the fillet be $r = \frac{1}{8}$ in. and the material cold-rolled SAE 1015. Determine the dimensions t, h ($b = 1.3h$) for a design factor of 1.8 based on variable stresses. Consider sections at A and B, indefinite life.

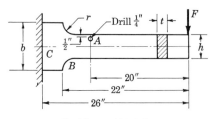

Problems 168-169.

169. The same as **168**, except that the load F varies from -1000 lb. to $+5000$ lb.

170. The beam shown is made of AISI C1020 steel, as rolled; $e = 8$ in. The load F is repeated from zero to a maximum of 1400 lb. Assume that the stress concentration at the point of application of F is not decisive. Determine the depth h and width t if $h \approx 4t$; $N = 1.5 \pm 0.1$ for Soderberg's line. Iteration is necessary because K_f depends on the dimensions. Start by assuming a logical K_f for a logical h (Fig. AF 11), with a final check of K_f. Considerable estimation inevitable.

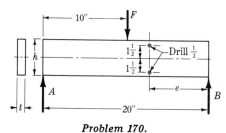

Problem 170.

171. Design a crank similar to that shown with a design factor of 1.60 ± 0.16 based on the modified Goodman line. The crank is to be forged with certain surfaces milled as shown and two $\frac{1}{4}$-in. holes. It is estimated that the material must be of the order of AISI 8630, WQT 1100°F. The length $L = 17$ in., $a = 5$ in.,

and the load varies from $+15$ kips to -9 kips. (a) Compute the dimensions at section AB with $h = 3b$. Check the safety of the edges (forged surfaces). (Iteration involved; one could first make calculations for forged surfaces and then check safety at holes.) (b) Without redesigning but otherwise considering relevant factors quantitatively, discuss actions that might be taken to reduce the size; holes must remain as located.

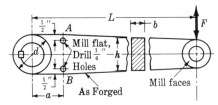

Problems 171-174.

172. The same as **171**, except that the Soderberg line is used.

CHECK PROBLEMS

173. For the crank shown, $L = 15$ in., $a = 3$ in., $d = 4.5$ in., $b = 1.5$ in. It is as forged from AISI 8630, WQT 1100°F, except for machined areas indicated. The load F varies from $+5$ kips to -3 kips. The crank has been designed without detailed attention to factors that affect its endurance strength. In section AB only, compute the factor of safety by the Soderberg criterion. Suppose it were desired to improve the margin of safety, with significant changes of dimensions prohibited, what various steps could be taken? What are your particular recommendations?

174. The same as **173**, except that after further study it is decided that the cycles of loading would not exceed 3×10^5. Compare results from the different equations given in the *Text* (§ 4.16).

175. The link shown is made of AISI C1020, as rolled, machined all over. It is loaded in tension by pins in the $D = \frac{3}{8}$-in. holes in the ends; $a = \frac{9}{16}$ in., $t = \frac{5}{16}$ in., $b = \frac{9}{16}$ in. $h = 1\frac{1}{8}$ in. Considering sections at A, B, and C, determine the maximum safe axial load for $N = 2$ and indefinite life (a) if it is repeated and reversed; (b) if it is repeated, varying from zero to maximum; (c) if it repeatedly varies from $F = -W$ to $F = 3W$. (d) Using the results from (a) and (b), determine the ratio of the endurance strength for a repeated load to that for a reversed load (Soderberg line).

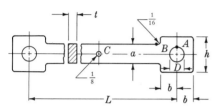

Problems 175-178.

176. The same as **175**, except that the material is magnesium alloy, AZ 80A-T5; 5×10^8 cycles.

177. The same as **175**, except that the life is to be 10^5 cycles.

178. The link shown has a vertical load applied at its center C and the fit of the pins is such that it acts as a simple beam; $L = 8$ in. and other dimensions are given in **175**. The material is stainless steel 410, OQT 850°F. For $N = 2$, determine the safe load F if it is (a) repeated and reversed, (b) repeatedly varying from zero to F, and (c) repeatedly varying from $F = -W$ to $F = 3W$. Consider points at B and C.

179. A steel rod as shown, AISI 2320, hot rolled, has been machined to the following dimensions: $D = 1$ in., $c = \frac{3}{4}$ in., $e = \frac{1}{8}$ in. A semicircular groove at the midsection has $r = \frac{1}{8}$ in.; for the radial hole, $a = \frac{1}{4}$ in. An axial load of 5 kips is repeated and reversed ($M = 0$). Compute the factor of safety (Soderberg) and make a judgment on its suitability (consider statistical variations of endurance strength—§ 4.4). What steps may be taken to improve the design factor?

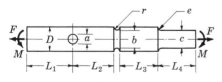

Problems 179-183.

180. The same as **179**, except that the material is AISI 1144 ETD (elevated temperature drawn—Table AT 10).

181. The same as **179**, except that the cycles of loading will not exceed 2×10^5.

182. The dimensions and material are as given in **179**, but $F = 0$ and moments M are applied at each end. What may be the repeated moment M, zero to maximum, for $N = 1.4$? What is the maximum stress induced by this moment?

183. The dimensions and material are as given in **179**, but $F = 0$, $M = 0$, and the member is subjected to torque couples at each end. What may be the repeated torque, zero to a maximum, for $N = 1.4$? What is the corresponding maximum stress?

184. A cantilever beam is subjected to a repeated reversed load, $F = 5000$ lb. Because of space requirements the dimensions shown are $h = 4$ in., $t = 2$ in., $r = \frac{1}{4}$ in., $b = 4\frac{1}{2}$ in. (a) Decide upon an appropriate material and heat treatment by assuming one and computing N; consider only sections A and B; indefinite life. (b) If later decisions suggest that the number of loading cycles with the maximum F will not exceed 10^5, would a less expensive material serve? If so, which one?

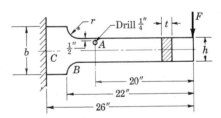

Problems 184, 185.

185. The same as **184**, except that F varies from -3 kips to $+9$ kips.

186. A stock stud that supports a roller follower on a needle bearing for a cam is made as shown, where $a = \frac{5}{8}$ in., $b = \frac{7}{16}$ in., $c = \frac{3}{4}$ in. The nature of the junction of the diameters at B is not defined. Assume that the inside corner is sharp. The material of the stud is AISI 2317, OQT 1000°F. Estimate the safe, repeated load F for $N = 2$. The radial capacity of the needle bearing is given as 1170 lb. at 2000 rpm for a 2500-hr. life. See Fig. 20.9, p. 532, *Text*.

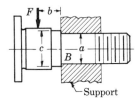

Problem 186.

187. The link shown is made of AISI C1035 steel, as rolled, with the following dimensions: $a = \frac{3}{8}$ in., $b = \frac{7}{8}$ in., $c = 1$ in., $d = \frac{1}{2}$ in., $L = 12$ in., $r = \frac{1}{16}$ in. The axial load F varies from 3000 lb. to 5000 lb. and is applied by pins in the holes. (a) What are the factors of safety at points A, B, and C if the link is machined all over? (b) What are the maximum stresses at these points?

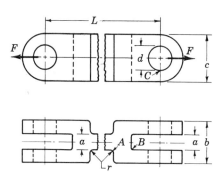

Problems 187, 188.

188. The same as **187**, except that the loading is cycled 200,000 times.

IMPACT PROBLEMS

189. A wrought-iron bar is 1 in. in diameter and 5 ft. long. (a) What will be the stress and elongation if the bar supports a static load of 5000 lb? Compute the stress and elongation if a 5000 lb. weight falls freely 0.05 in. and strikes a stop at the end of the bar. (b) The same as (a), except that the bar is aluminum alloy 3003-H14.

190. What should be the diameter of a rod 5 ft. long, made of an aluminum alloy 2024-T4, if it is to resist the impact of a weight of $W = 500$ lb. dropped through a distance of 2 in.? The maximum computed stress is to be 20 ksi. Compare solution with that in § 4.37, *Text*, and state the reasons for a different answer.

191. A rock drill has the heads of the cylinder bolted on by $\frac{7}{8}$-in. bolts somewhat as shown. The grip of the bolt is 4 in. (a) If the shank of the bolt is turned down to the minor diameter of the coarse-thread screw, 0.7387 in., what energy may each bolt absorb if the stress is not to exceed 25 kips? (b) Short bolts used as described above sometimes fail under repeated shock loads. It was found in one instance that if long bolts, running from head to head, were used, service failures were eliminated. How much more energy will the bolt 21 in. long absorb for a stress of 25 ksi than the bolt 4 in. long? As before, let the bolt be turned down to the minor diameter. The effect of the threads on the strength is to be neglected.

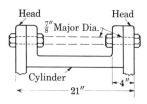

Problem 191.

192. As seen in the figure, an 8.05-lb. body A is moving down with a constant acceleration of 12 fps², having started from rest at point C. If A is attached to a steel wire, W&M gage 8 (0.162 in. diameter) and if for some reason

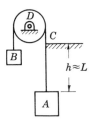

Problems 192, 193.

the sheave D is instantly stopped, what stress is induced in the wire?

193. The hoist A shown, weighing 5000 lb. and moving at a constant $v = 4$ fps is attached to a 2 in. wire rope that has a metal area of 1.6 sq. in. and a modulus $E = 12 \times 10^6$ psi. When $h = 100$ ft., the sheave D is instantly stopped by a brake (since this is impossible, it represents the worst conceivable condition). Assuming that the stretching is elastic, compute the maximum stress in the rope.

194. A coarse-thread steel bolt, $\frac{3}{4}$ in. in diameter, with 2 in. of threaded and 3 in. of unthreaded shank, receives an impact caused by a falling 500-lb. weight. The area at the root of the threads is 0.334 sq. in. and the effects of threads are to be neglected. (a) What amount of energy in in-lb. could be absorbed if the maximum calculated stress is 10 ksi? (b) From what distance h could the weight be dropped for this maximum stress? (c) How much energy could be absorbed at the same maximum stress if the unthreaded shank were turned down to the root diameter?

195. The same as **194**, except that annealed aluminum bronze is used for its corrosion-resistance property and strength.

196. A part of a machine that weighs 1000 lb. is raised and lowered by a $1\frac{1}{2}$-in. steel rod that has Acme threads on one end (see § 8.18, *Text*, for minor diameter). The length of the rod is 10 ft. and the upper 4 ft. are threaded. As the part is being lowered it sticks, then falls freely a distance of $\frac{1}{8}$ in. (a) Compute the maximum stress in the rod. (b) What would be the maximum stress in the rod if the lower end had been turned down to the root diameter?

197. A weight W of 50 lb. is moving on a smooth horizontal surface with a velocity of 2 fps when it strikes head-on the end of a $\frac{3}{4}$ in. round steel rod, 6 ft. long. Compute the maximum stress in the rod. What design factor based on yield strength is indicated for AISI 1010, cold drawn?

198. The same as **197**, except that the coefficient of friction between the weight and the surface on which it slides is 0.2.

199. A rigid weight of 100 lb. is dropped a distance of 25 in. upon the center of a 12 in., 50-lb. I-beam ($I_x = 301.6$ in.⁴) that is simply supported on supports 10 ft. apart. Compute the maximum stress in the I-beam both with and without allowing for the beam's weight.

200. The same as **199**, except that the beam is a cantilever and the load strikes the end. Is the stress within the elastic limit for structural steel?

201. A 3000 lb. automobile (here considered rigid) strikes the midpoint of a guard rail that is an 8-in., 23-lb. I-beam, 40 ft. long; $I = 64.2$ in.⁴. Made of AISI C1020, as rolled, the I-beam is simply supported on rigid posts at its ends. (a) What level velocity of the automobile results in stressing the I-beam to the tensile yield strength? Compare results obtained by including and neglecting the beam's mass.

DATA LACKING—DESIGNER'S DECISIONS

202. A simple beam is struck midway between supports by a 32.2-lb. weight that has fallen 20 in. The length of the beam is 12 ft. If the stress is not to exceed 20 ksi, what size I-beam should be used?

203. The same as **202**, except that the beam is a cantilever and weight strikes the free end.

204. A 10-in., 25.4-lb. I-beam, AISI 1020, as rolled, is 10 ft. long and is simply supported at the ends as shown. There is a static load of $F_1 = 10$ kips, 4 ft. from the left end, and a repeated reversed load of $F_2 = 10$ kips, 3 ft. from the right end. It is desired to make two attachments to

Problem 204.

the beam through holes as shown. No significant load is supported by these attachments, but the holes cause stress concentration. Will it be safe to make these attachments as planned? Determine the factor of safety at the point of maximum moment and at points of stress concentration.

205. The runway of a crane consists of $L = 20$-ft. lengths of 15-in., 42.9-lb. I-beams, as shown, each section being supported at its ends; AISI C1020, as rolled. The wheels of the crane are $a = 9$ ft. apart, and the maximum load expected is $F = 10,000$ lb. on each wheel. Neglecting the weight of the beam, find the design factor (a) based on variable stresses for 10^5 cycles, (b) based on the ultimate strength. (HINT. Since the maximum moment will occur under a wheel, assume the wheels at some distance x from the point of support, and determine the reaction R_1 as a function of x; $dM/dx = 0$ gives position for a maximum bending moment.)

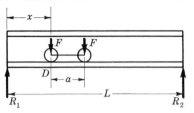

Problem 205.

206. If you have worked one of the union-of-rods problems **49–68,** allow for the factors discussed in this chapter and compute the factor of safety based on variable stresses. Sections that might be considered include; at the collar, both sides; at the inside corner of the slot in the socket; at the bottom of the hole inside the socket. Do these calculations indicate good design? Redesign as necessary if so advised by the instructor.

207–220. These numbers may be used for other problems.

Section 3

SCREW FASTENINGS

SIMPLE TENSION INCLUDING TIGHTENING STRESSES

DESIGN PROBLEMS

221. A 5000-lb. gear box is provided with a steel (as-rolled B1113) eye bolt for use in moving it. What size bolt should be used: (a) if UNC threads are used? (b) if UNF threads are used? (c) if the 8-thread series is used? Explain the basis of your choice of design factor.

222. A motor weighing 2 tons is lifted by a wrought-iron eye bolt which is screwed into the frame. Decide upon a design factor and determine the size of the eye bolt if (a) UNC threads are used, (b) UNF threads are used. *Note:* Fine threads are not recommended for brittle materials.

223. The same as **222**, except that the motor weighs $1\frac{1}{2}$ tons.

224. A wall bracket, Fig. 8.13, *Text,* is loaded so that the two top bolts that fasten it to the wall are each subjected to a tensile load of 710 lb. The bolts are to be cold forged from AISI C1020 steel with UNC threads. Neglecting the effect of shearing stresses, determine the diameter of these bolts if they are well tightened.

225. A connection similar to Fig. 5.9, *Text,* is subjected to an external load F_e of 1250 lb. The bolt is made from cold-finished AISI B1113 steel with UNC threads. (a) Determine the diameter of the bolt if it is well tightened. (b) Compute the initial tension and the corresponding approximate tightening torque if $s_i = 0.85 s_y$ (§ 5.8).

226. The cylinder head of a 10 × 18 in. Freon compressor is attached by 10 stud bolts made of SAE grade 5. The cylinder pressure is 200 psi. (a) What size bolts should be used? (b) What approximate tightening torque should be needed to induce a tightening stress s_i of 0.9 times the proof stress?

227. The American Steel Flange Standard specifies that 8 bolts are to be used on flanges for 4 in. pipe where the steam or water pressure is 1500 psi. It is also specified that, in calculating the bolt load, the outside diameter of the gasket, which is $6\frac{3}{16}$ in., should be used. Determine (a) the diameter of the UNC bolts if they are well tightened and made of ASTM 354 BD (Table 5.2), (b) the approximate torque to tighten the nuts if the initial stress is 90% of the proof stress. The Standard specifies that $1\frac{1}{4}$-in. bolts with 8 th./in. be used (these bolts are also subjected to bending). How does your answer compare?

CHECK PROBLEMS

228. A cap screw, $\frac{3}{4}$ in.-10-UNC-2, with a hexagonal head that is $\frac{9}{16}$ in. thick, carries a tensile load of 3000 lb. If the material is AISI 1015, cold drawn, find the factor of safety based on ultimate strengths of (a) the threaded shank, (b) the head against being sheared off, and (c) the bearing surface under the head. (d) Is there any need to consider the strength of standard cap-screw heads in design?

229. A bolt, $1\frac{1}{8}$ in.-7-UNC-2, is subjected to a tensile load of 10,000 lb. The head has a thickness of $\frac{3}{4}$ in. and the nut a thickness of 1 in. If the material is SAE grade 2 (Table 5.2), find the design factor as based on ultimate stresses (a) of the threaded shank, (b) of the head against being sheared off, and (c) of the bearing surface under the head. The bolt head is finished. (d) Is there any need to consider the strength of standard bolt heads in design?

230. An axial force is applied to a regular nut which of course tends to shear the threads on the screw. (a) What is the ratio of the force necessary to shear the threads (all threads initially in intimate contact) to the force necessary to pull the bolt in two? Use coarse threads, a $1\frac{1}{2}$-in. bolt, and assume that $s_{us} = 0.75s_u$. The head thickness is 1 in. and the nut thickness is $1\frac{5}{16}$ in. (b) Is failure of the thread by shear likely in this bolt?

231. For bolted structural joints, specifications suggest that $\frac{1}{2}$-in. bolts (high-strength material) be tightened to an initial tension of $F_i = 12,500$ lb. What should be the approximate tightening torque? How does your answer compare with $T = 90$ ft-lb., which is the value in the specification?

232. One method of estimating the initial tensile stress in a tightened bolt is to turn the nut until it is snug, but with no significant stress in the bolt. Then the nut is turned through a predetermined angle that induces a certain unit strain corresponding to the desired stress. A $\frac{3}{4}$-in. bolt of the type shown in Fig. 5.4,

Text, is turned down until, for practical purposes, the diameter of the entire shank is the minor diameter. The material is AISI 4140, OQT 1200°F. The grip is 5 in. and the effective strain length is estimated to be 5.3 in. If the initial tensile stress at the root diameter is to be about 75% of the yield strength, through what angle should the nut be turned after it is just snug? The threads are UNC and the parts being bolted are assumed to be rigid.

233. The same as **232**, except that the threads are UNF.

234. When both ends of a bolt are accessible for micrometer measurements, the total elongation δ caused by tightening can be determined by measuring the lengths before and after tightening. In order to reduce this total elongation to unit elongation, thence to stress, the effective strain length for the bolt must be known. For a $1\frac{1}{4}$-in. steel bolt, threaded for its full length, 8-thread series, the effective strain length has been found by experiment to be $L_e = 0.97G + 1.1$ in., where G is the grip (by W. A. McDonald, North Carolina State College). Let the bolt material be AISI 8742, OQT 1000°F. (a) It is desired that the initial tensile stress be about $0.7s_y$. What total elongation should be obtained for a grip length of 4.8 in.? (b) Investigate the approximate tightening torque for the specified condition. How could this torque be obtained?

ELASTIC CONSIDERATIONS

235. The member C shown is part of a swivel connection that is to be clamped by a 1-in. bolt D to the member B, which has large dimensions in the plane perpendicular to the paper. Both B and C are aluminum alloy 2024-T4, HT aged. The bolt is made of AISI C1113, cold-drawn steel; consider the unthreaded shank to be 2 in. long; it is well tightened with a torque of 250 ft-lb.; UNC threads, unlubricated. (a) Estimate the initial tension by equation (5.2), assume elastic action, and compute the total bolt elongation and the total deformation of B and C. Let the effective strain length be 2 in. (b) After tightening, an external axial force F_e of 5000 lb. is applied to member C. Determine the total normal stresses in the bolt and in B and C. (c) Determine the load required to "open" the connection. Draw a diagram similar to Fig. 5.6, *Text*, locating points A, B, D and M.

rigid. (b) What are the stresses if an axial load of 5 kips is now applied to the bolt ends? (c) Compute the bolt load that just results in a zero stress in the tube.

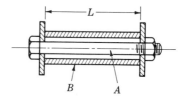

Problem 238.

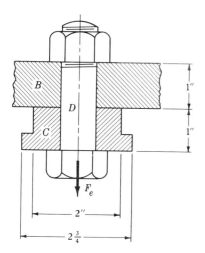

Problems 235, 236.

236. The same as **235**, except that the materials are reversed, the bolt is aluminum and B and C are steel.

237. A 1-in. steel bolt is used to clamp two aluminum (2014-T6, HT aged) plates together as shown by Fig. 5.9, *Text*. The aluminum plates have a total thickness of 2 in. and an equivalent diameter of 2 in. The bolt is heated to a temperature of 200°F, then inserted in the aluminum plates, which are at 80°F, and tightened so as to have a tensile tightening stress of 30 ksi in the unthreaded shank while still at 200°F. What is the tensile stress in the bolt after the assembly has cooled to 80°F? The deformations are elastic.

238. A $1\frac{1}{8}$-in. steel bolt A passes through a yellow brass (B36-8) tube B as shown. The length of the tube is 30 in. (virtually the unthreaded bolt length), the threads on the bolt are UNC, and the tube's cross-sectional area is 2 sq. in. After the nut is snug it is tightened $\frac{1}{4}$ turn. (a) What normal stresses will be produced in the bolt and in the tube? Assume that washers, nut, and head are

ENDURANCE STRENGTH

DESIGN PROBLEMS

239. As shown diagrammatically, a bearing is supported in a pillow block attached to an overhead beam by two cap screws, each of which, it may be assumed, carries half the total bearing load. This load acts vertically downward, varying from 0 to 1500 lb. The screws are to be made of AISI C1118, as rolled, and they are tightened to give an initial stress of about $s_i = 0.5s_y$. The pillow block is made of class-20 cast iron. Assume that the effective length of screw is equal to the thickness t, as shown, and that the head and beam are rigid (overly conservative?). The equivalent diameter of the compression area may be taken as twice the bolt diameter. For a design factor of 1.75, determine the size of the screw: (a) from the Soderberg line, (b) from the modified Goodman line. (c) What size do you recommend using?

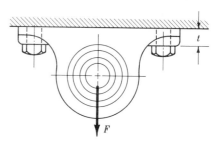

Problem 239.

240. A connection similar to Fig. 5.9, *Text*, is subjected to an external load that varies from 0 to 1250 lb. The bolt is cold forged from AISI B1113 steel; UNC threads. The aluminum parts C (3003 H14) have a total thickness of $1\frac{1}{2}$ in. and an external diameter of $2D$. It is desired that the connection not open for an external load of $1.5F_e$. Determine (a) the initial tensile load on the bolt, (b) the bolt diameter for $N = 2$ based on the Soderberg line. (c) Compare with the size from **225** and discuss the difference, if any.

241. The same as **240**, except that the materials are reversed; parts C are steel, bolts are aluminum.

242. The same as **240**, except that the parts C are made of magnesium alloy AZ80A.

243. This problem concerns the Freon compressor of **226**: size, 10×18 in.; 10 studs, UNC; made of C1118, as rolled; 200 psi gas pressure. The initial tension in the bolts, assumed to be equally loaded, is such that a cylinder pressure of 300 psi is required for the joint to be on the point of opening. The bolted parts are cast steel and for the first calculations, it will be satisfactory to assume the equivalent diameter of the compressed parts to be twice the bolt size. (a) For $N = 2$ on the Soderberg criterion, what bolt size is required? (b) Compute the torque required for the specified initial tension.

244. The same as **243**, except that the pressure to open the joint is 400 psi.

245. A cast-iron (class 35) Diesel-engine cylinder head is held on by 8 stud bolts with UNC threads. These bolts are made of AISI 3140 steel, OQT 1000°F (Fig. AF2). Assume that the compressed material has an equivalent diameter twice the bolt size. The maximum cylinder pressure is 750 psi and the bore of the engine is 8 in. Let the initial bolt load be such that a cylinder pressure of 1500 psi brings the joint to the point of opening. For a design factor of 2, determine the bolt diameter (a) using the Soderberg equation, (b) using the Goodman equation. (c) What approximate torque will be required to induce the desired initial stress? (d) Determine the ratio of the

initial stress to the yield strength. Considering the lessons of experience (§ 5.8), what initial stress would you recommend? Using this value, what factor of safety is computed from the Soderberg equation?

246. A 30,000-lb. body is to be mounted on a shaker (vibrator). The shaker will exert a harmonic force of $F = 30,000 \sin 2\pi t f$ lb. on the body where f cps is the frequency and t sec. is the time. The frequency can be varied from 5 to 10,000 cps. The harmonic force will exert a tensile load on the bolts that attach the body to the shaker when F is positive. Determine the minimum number of $\frac{1}{2}$-in.-UNF bolts that must be used for $N = 2$ based on the Soderberg line. The material of the bolts is to be AISI 8630, WQT 1100°F; the material of the body that is to be vibrated is aluminum alloy, 2014-T6 and the joint is not to open for an external force that is 1.25 times the maximum force exerted by the shaker. It may be assumed that the equivalent diameter of the material in compression is twice the bolt diameter.

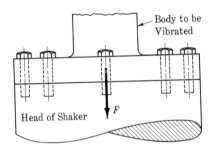

Problems 246, 247.

247. The same as **246**, except that the diameter of the bolt is to be 1-in.

248. The maximum external load on the cap bolts of an automotive connecting rod end, imposed by inertia forces at top dead center, is taken to be 4000 lb.; the minimum load is zero at bottom dead center. The material is AISI 4140, OQT 1100°F (qualifying for SAE grade 5); assume that $s_n' = 0.45s_u$. The grip for through bolts is 1.5 in. For design purposes, let each bolt take half the load, and use an equivalent $D_e = 1\frac{3}{8}$ in. for the connected parts. The threads extend a negligible amount into the grip. For the initial computation, use an opening load $F_0 = 1.75F_e$. Considering the manner in which the bolt is loaded, we decide that a design factor of 1.4 (Soderberg) should be quite adequate. (a) Does a $\frac{5}{16}$-24 UNF satisfy this situation? If not, what size do you recommend? (b) Experience suggests that, in situations such as this, an initial stress of the order suggested in § 5.8, *Text*, is good insurance against fatigue failure. Decide upon such an s_i and recompute N. How does it change? Would you be concerned about the safety in this case? Consider the variation of s_i as a consequence of the use of a torque wrench and also the stress relaxation with time (due to seating and other factors), and discuss. (c) Compute the required tightening torque for each s_i.

CHECK PROBLEMS

249. A 1-in. steel bolt A (normalized AISI 1137, cold-rolled threads) passes through a yellow brass tube B (B36-8, $\frac{1}{2}$ hard) as shown. The tube length is 30 in., its cross-sectional area is 2 sq. in., and the UNC bolt threads extend a negligible amount below the nut. The steel washers are $\frac{1}{4}$ in. thick and are assumed not to bend (clearances are exaggerated). The nut is turned snug and then tightened $\frac{1}{4}$ turn. (a) If an external tensile axial load, varying from 0 to 5 kips, is repeatedly applied to the bolt what is the factor of safety of the bolt by the Soderberg criterion? (b) What is the external load on the bolt at the instant that the load on the tube becomes zero?

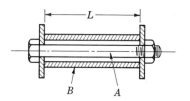

Problems 249, 250.

250. A $\frac{3}{4}$-in. fine-thread bolt, made of AISI 1117, cold drawn, with rolled threads, passes through a yellow brass tube and two steel washers, as shown. The tube is 4 in. long, $\frac{7}{8}$-in. internal diameter, $1\frac{1}{4}$-in. external diameter. The washers are each $\frac{1}{4}$ in. thick. The unthreaded part of the bolt is 3 in. long.

Assume that there is no stretching of the bolt inside of the nut in finding its k. The unlubricated bolt is tightened by a torque of 1800 in-lb. The external load, varying from 0 to 4 kips, is axially applied to the washers an indefinite number of times. (a) Compute the factor of safety of the bolt by the Soderberg criterion. Is there any danger of failure of the bolt? (b) What pull must be exerted by the washers to remove all load from the brass tube?

251. A coupling bolt (§ 5.13, *Text*) of ASTM 354 BC (Table 5.2, *Text*) is used to connect two parts made of cast iron, class 35. The diameter of the coarse-thread bolt is $\frac{1}{2}$ in.; its grip is 2 in., which is also nearly the unthreaded length. The bolt is tightened to have an initial tension of 4000 lb. The parts support an external load F_e that tends to separate them, and it varies from zero to 5000 lb. What is the factor of safety (Soderberg)?

252. The same as **251**, except that the connecting parts are separated by a vellumoid gasket.

253. The cap on the end of a connecting rod (automotive engine) is held on by two $\frac{5}{16}$-in. bolts that are forged integrally with the main connecting rod. These bolts have UNF threads with a $\frac{5}{8}$-in. grip on an unthreaded length of virtually $\frac{5}{8}$ in. The nuts are to be tightened with a torque of 20 ft-lb. and the maximum external load on one bolt is expected to be 2330 lb. Let the equivalent diameter of the connected parts be $\frac{3}{4}$ in. (a) Estimate the maximum force on the bolt. (b) Compute the opening load. Is this satisfactory? (c) If the bolt material is AISI 4140, OQT 1000°F, what is the factor of safety based on the Soderberg criterion?

SET SCREWS

254. A 6-in. pulley is fastened to a $1\frac{1}{4}$-in. shaft by a set screw. If a net tangential force of 75 lb. is applied to the surface of the pulley, what size screw should be used when the load is steady?

255. An eccentric is to be connected to a 3-in. shaft by a set screw. The center of the eccentric is $1\frac{1}{4}$ in. from the center of the shaft when a tensile force of 1000 lb. is applied to the eccentric rod perpendicular to the line of centers. What size set screw should be used for a design factor of 6?

256. A lever 16 in. long is to be fastened to a 2-in. shaft. A load of 40 lb. is to be applied normal to the lever at its end. What size of set screw should be used for a design factor of 5?

257. A 12-in. gear is mounted on a 2-in. shaft and is held in place by a $\frac{7}{16}$-in. set screw. For a design factor of 3, what would be the tangential load that could be applied to the teeth and what horsepower could be transmitted by the screw?

258–270. These numbers may be used for other problems.

Section 4

SPRINGS

HELICAL COMPRESSION SPRINGS

DESIGN—LIGHT, MEDIUM SERVICE

271. A solenoid brake (Fig. 18.2, *Text*) is to be actuated by a helical compression spring. The spring should have a free length of approximately 18 in. and is to exert a maximum force of 2850 lb. when compressed to a length of 15 in. The outside diameter must not exceed 7 in. Using oil-tempered wire, design a spring for this brake, (wire diameter, coil diameter, number of active coils, pitch, pitch angle, "solid stress"). General Electric used a spring made of 1 in. wire, with an outside diameter of 6 in., and 11½ free coils for a similar application.

272. A coil spring is to be used for the front spring of an automobile. The spring is to have a rate of 400 lb./in., an inside diameter of 4³⁄₆₄ in., and a free length of 14⅛ in., with squared-and-ground ends. The material is to be oil-tempered chrome-vanadium steel. Decide upon the diameter of the wire and the number of free coils for a design load of $F = 1500$ lb. Be sure "solid stress" is all right. How much is the pitch angle?

273. A coiled compression spring is to fit inside a cylinder ⅝ in. in diameter. For one position of the piston, the spring is to exert a pressure on the piston equivalent to 5 psi of piston area, and in this position, the overall length of the spring must not exceed (but may be less than) 2 in. A pressure of 46 psi on the piston is to compress the spring ¾ in. from the position described above. Design a spring for medium service. Specify the cheapest suitable material, number of total and active coils for squared-and-ground ends, and investigate the pitch angle, and "solid stress."

274. A helical spring is to fit about a 1¹⁄₁₆-in. rod with a free length of 2¾ in. or less. A maximum load of 8 lb. is to produce a deflection of 1¾ in. The spring is expected to be compressed less than 5000 times during its life, but is subjected to relatively high temperatures and corrosive atmosphere. Select a material and determine the necessary wire size, mean coil diameter, and number of coils. Meet all conditions advised by *Text*.

275. In order to isolate vibrations, helical compression springs are used to support a machine. The static load on each spring is 3500 lb., under which the deflection should be about 0.5 in. The "solid deflection" should be about 1 in. and the outside coil diameter should not exceed 6 in. Recommend a spring for this application; include scale, wire size, static stress, material, number of coils, solid stress, and pitch of coils.

CHECK PROBLEMS—LIGHT, MEDIUM SERVICE

276. The front spring of an automobile has a total of $9\frac{1}{2}$ coils, $7\frac{3}{8}$ active coils (squared-and-ground ends), an inside diameter of $4\frac{3}{64}$ in., and a free length of $14\frac{1}{4}$ in. It is made of SAE 9255 steel wire, OQT 1000°F, with a diameter of $\frac{43}{64}$ in. Compute (a) the rate (scale) of the spring; (b) the "solid stress" and compare with a permissible value (is a stop needed to prevent solid compression?). (c) Can 95% of the solid stress be repeated 10^5 times without danger of failure? Would you advise shot peening the spring?

277. An oil-tempered steel helical compression spring has a wire size of No. 3 W&M, a spring index of 4.13, 30 active coils, a pitch of 0.317 in., ground-and-squared ends; medium service. (a) What maximum load is permitted if the recommended stress is not exceeded (static approach)? Compute (b) the corresponding deflection, (c) "solid stress," (d) pitch angle, (e) scale, (f) the energy absorbed by the spring from a deflection of 0.25 in. to that of the working load. (g) Is there any danger of this spring buckling? (h) What maximum load could be used if the spring were shot peened?

278. The same as **277**, except that the material is phosphor bronze.

279. The same as **277**, except that the material is spring brass.

280. It is desired to isolate a furnace, weighing 47,300 lb., from the surroundings by mounting it on helical springs. Under the weight, the springs should deflect approximately 1 in., and at least 2 in. before becoming solid. It has been decided to use springs having a wire diameter of 1 in., an outside diameter of $5\frac{3}{8}$ in., 4.3 free coils. Determine (a) the number of springs to be used, (b) the stress caused by the weight, (c) the "solid stress." (d) What steel should be used?

281. Using the "static approach" (§ 6.9, *Text*), let the stress from equation (6.1) equal the design stress from column 3, Table AT 17, which can be written as $s_{sd} = Q/D_w{}^x$, and derive the equation

$$\frac{D_m{}^{2-x}}{F} = \frac{8KC^{3-x}}{\pi Q}.$$

(b) Plot a curve of $D_m{}^{2-x}/F$ vs. C (from $C = 3$ to 10) for oil-tempered wire (ASTM A 229); light service. Note that given D_m and F, it is then possible to determine C and D_w without a trial-and-error solution. (See Hirshorn, "Helical Spring Design," *Machine Design*, February 19, 1959, for additional curves).

VARYING STRESS APPROACH

DESIGN PROBLEMS

282. A spring, subjected to a load varying from 100 lb. to 250 lb., is to be made of oil-tempered, cold-wound wire. Determine the diameter of the wire and the mean diameter of the coil for a design factor of 1.25 based on Wahl's line. The spring index is to be at least 5. Conform to good practice, showing checks for all significant parameters. Let the free length be between 6 and 8 in.

283. A carbon-steel spring is to be subjected to a load that varies from 500 to 1200 lb. The outside diameter should be between 3.5 and 4 in., the spring index between 5 to 10; approximate scale of 500 lb./in. Choose a steel, and for a design factor of 1.4 by the Wahl line, find the wire diameter. Also determine the number of active coils and the free length for squared-and-ground ends. Conform to the general conditions specified in the *Text*.

284. A helical compression spring, made of oil-tempered, cold-wound carbon steel, is to be subjected to a working load varying from 100 to 300 lb. for an indefinite time (severe). A mean coil diameter of 2 in. should be satisfactory. (a) Using the static approach, compute a wire diameter. (b) For this wire size, compute the factor of safety as given by the Wahl line. (c) For satisfactory D_w, decide upon details (pitch angle, "solid stress," scale).

285. A helical spring of hard-drawn wire with a mean diameter of $1\frac{1}{2}$ in. and square-and-ground ends is to be subjected to a maximum load of 325 lb. (a) Compute the wire diameter for average service. (b) How many total coils are required if the scale is 800 lb./in.? (c) For a minimum load of 100 lb., what is the factor of safety according to the Wahl line? Would it be safe for an indefinite life?

286. A helical spring is to be subjected to a maximum load of 200 lb. (a) Determine the wire size suitable for medium service if the material is carbon steel ASTM H230; $C = 6$. Determine the factor of safety of this spring according to the Wahl line (b) if the minimum force is 150 lb., (c) if the minimum force is 100 lb., (d) if the minimum force is 25 lb.

287. The same as **286,** except design for light service.

288. The same as **286,** except design for severe service.

CHECK PROBLEMS

289. A Diesel valve spring is made of $\frac{3}{8}$-in. chrome-vanadium steel wire, shot peened; inside diameter is 3 in., 7 active coils, free length is $7\frac{3}{8}$ in., solid length is $4\frac{1}{8}$ in., length with valve closed, $6\frac{1}{4}$ in., length when open, $5\frac{1}{8}$ in. (a) Compute the spring constant and the factor of safety as defined by the Wahl criterion (see § 6.13, *Text*). (b) Is there any danger of damage to the spring if it is compressed solid? (c) What is the natural frequency? If this spring is used on a 4-stroke Diesel engine at 450 rpm, is there any danger of surge? (d) Compute the change of stored energy between working lengths.

290. A Diesel valve spring is made of $\frac{5}{16}$-in. chrome-vanadium steel wire, shot peened; inside diameter is $2\frac{1}{16}$ in., 10 active coils; free length, $6\frac{1}{2}$ in., solid length, $3\frac{3}{4}$ in., length when closed, $5\frac{1}{2}$ in., when open, $4\frac{3}{8}$ in. Parts (a)–(d) are the same as in **289.**

291. A helical spring is hot wound from $\frac{5}{8}$-in. carbon-steel wire with an outside diameter of $3\frac{1}{4}$ in. A force of 3060 lb. is required to compress the spring $1\frac{3}{4}$ in. to the solid height. In service the spring is compressed so that its deformation varies from $\frac{1}{2}$ in. to $1\frac{1}{8}$ in. (a) What is the factor of safety by the Wahl criterion? (b) Is the "solid stress" safe?

Compute (c) the pitch angle, (d) the change of stored energy between the working lengths, (e) the factor of safety if the spring is peened?

ENERGY STORAGE

292. Derive equation (j) of § 6.7.

293. A 10-lb. body falls 10 in. and then strikes a helical spring. Design a hard-drawn carbon steel spring that will absorb this shock occasionally without permanent damage. Determine appropriate values of wire diameter, coil diameter, pitch, free length, closed length, and the maximum stress under the specified conditions, and scale. Let $C = 7$.

294. A helical spring, of hard-drawn steel wire, is to absorb 75 in-lb. of energy without being stressed beyond the recommended value for average service. Let $C = 6$. Decide upon satisfactory dimensions: D_w, D_m, N_c, free length, pitch angle, solid stress, volume of metal, possibility of the spring buckling. (INSTRUCTOR: This problem may be assigned as a class project so that each one of 4 or more groups completes a design for a certain combination of F and δ. For example, values of F of (a) 25 lb. (b) 50 lb. (c) 75 lb. (d) 100 lb. (e) 125 lb. Each member of each group should make an independent solution and then each group should decide on its best answer. Tabulate the results of all groups and make a choice.)

295. The same as **294,** except that the material is phosphor bronze.

296. The same as **294,** except that the material is chromium-vanadium steel.

CONCENTRIC HELICAL SPRINGS

297. Two concentric helical springs are to be subjected to a load that varies from a maximum of 235 lb. to a minimum of 50 lb. They are to fit inside a 1 in. cylinder. The maximum deflection is to be $\frac{3}{4}$ in., and the deflection when compressed solid is to be approximately 1 in. Using the "static approach" for severe service (maximum load), determine the wire diameter, mean coil diameter, number of coils, solid length, and free length of both springs. (Start with oil-tempered wire and assume a diametral clearance between the outer spring and the cylinder of $D_w/2$, assume a similar clearance be-tween springs. Search for a suitable spring index and wire size.)

298. Two concentric, helical com-pression springs are used on a freight car. The larger spring has an outside diameter of 7 in., a free length of $7\frac{1}{8}$ in., and is made of a $1\frac{3}{8}$ in. steel bar. The smaller spring has an outside diameter of $4\frac{1}{8}$ in., a free length of $6^{13}\!\!/_{16}$ in., and is made of $\frac{7}{8}$ in. steel bar. The solid height of each spring is $5\frac{1}{4}$ in. and the forces required to compress them solid are 15,530 lb. and 7,000 lb., respectively. The working load of the two springs is 11,350 lb. Deter-mine (a) the number of free coils in each spring, (b) the stress in each when com-pressed solid, (c) the stresses induced by the working load. Notice that the outer spring deflects $\frac{5}{16}$ in. before the inner one takes a load. (d) What energy is absorbed while changing deflection from that at the working load to that when the springs are compressed "solid"?

TORSION-BAR SPRINGS

299. A torsion bar similar to that shown is to be used for the front spring of an automobile. Its rate should be 400 lb./in. of deflection of the end of the arm which is $e = 10$ in. long. It is made of AISI 9261, OQT 900°F, and the maxi-mum repeated load is 1500 lb. perpendicu-lar to the center line of the arm. The support is such that bending of the bar is negligible. (a) Determine its diameter and length so that no permanent set occurs due to a 30% overload (limited by a stop). Use $s_{ys} = 0.6s_y$, but check with equation (c), § 6.3, *Text*, if appropriate. (b) Determine the factor of safety accord-ing to the Soderberg criterion if the load varies from 1200 lb. to 1500 lb.; minimum $r/d = 0.1$, $D/d = 3$. (c) The same as (b)

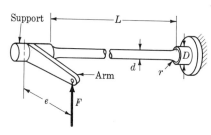

Problems 299, 300.

except that the bar is shot-peened all over. What other steps may be taken to improve the fatigue strength?

300. A solid steel torsion bar is loaded through a 10 in. arm as shown. The load F perpendicular to the center line of the arm varies from 500 to 1000 lb., 200,000 cycles. The bar is $d = \frac{7}{8}$ in. in diameter and 30 in. long; let $D/d = 3$; $r/d = 0.1$; (a) Determine the maximum shearing stress in the bar, the angular deflection, and the scale (lb./in.) where F is applied. The support is such that bending of the bar is negligible. (b) Select a material and heat treatment for this bar for a minimum $N = 2$, Soderberg criterion.

HELICAL SPRINGS—NONCIRCULAR SECTION

301. A spring is to be designed of square oil-tempered steel wire and sub-jected to a repeated maximum load of 325 lb.; mean coil diameter, $1\frac{1}{2}$ in.; deflection, $1^{3}\!\!/_{32}$ in. Determine (a) the wire size for average service, (b) the re-quired number of active coils, (c) the solid height, free length, and pitch (the ends are squared and ground, the "solid stress" must be satisfactory, and the pitch angle not excessive). (d) What amount of energy is stored when the load is 325 lb.? Express in in-lb. and Btu.

302. The same as **301**, except that the material is K-Monel.

303. A coil spring, of hard-drawn carbon steel, is to deflect 1 in. under a load of 100 lb. The outside coil diameter is to be 1 in. Compute the number of active coils, (a) if the wire is round, $\frac{5}{32}$ in. in diameter, (b) if the wire is square, $\frac{5}{32}$ in. on the side, (c) if the wire is rectangular $\frac{1}{8} \times \frac{3}{16}$ in., long dimen-sion parallel to the axis, (d) if the wire is rectangular $\frac{3}{16} \times \frac{1}{8}$ in., short dimension parallel to the axis. (e) What is the maxi-mum stress in each of the above springs under the 100-lb. load? (f) What is the ratio of the approximate volumes, square-or rectangular-wire to round-wire spring?

304. The same as **303**, except that the deflection is to be $1\frac{1}{2}$ in. under a load of 250 lb.; the outside diameter is to be $1\frac{1}{4}$ in.; only round and square wires are to be considered of size $\frac{3}{16}$ in.

TENSION SPRINGS

305. Design two tension springs for a spring balance with a capacity of 200 lb. Each spring supports a maximum load of 100 lb. The outside diameter must not exceed $1\frac{1}{4}$ in. and the total length including end loops must not exceed $9\frac{1}{2}$ in. Select a material and determine the dimensions, including wire diameter, number of coils, and free length.

306. Two helical tension springs are to be used in scales for weighing milk. The capacity of the scales is 30 lb., each spring carries 15 lb. with a deflection of $3\frac{9}{16}$ in. The springs are made of No. 14 W&M steel wire, outside diameter, $2\frac{9}{32}$ in. (a) How many coils should each spring have? (b) What is the maximum stress in the wire? What material should be used?

307. A tension spring for a gas-control lever is made of $D_w = 0.078$-in. steel wire; inside diameter, 0.609 in.; number of coils, 55; free length including end loops, $5\frac{9}{16}$ in. When the spring is extended to a length of $6\frac{5}{16}$ in., it must exert a force $5\frac{1}{2}$ lb.; it must extend to $9\frac{5}{16}$ in. without damage. Determine (a) the initial tension, (b) the stress in the spring caused by the initial tension (compare with the recommended maximum values), (c) the stress caused by the $5\frac{1}{2}$-lb. load, (d) the maximum stress. What material should be used? (e) What energy is absorbed from the point where the load is the initial tension until the spring's length is $6\frac{5}{16}$ in.? (*Data courtesy Worthington Corporation.*)

TORSION SPRINGS

308. Derive an equation for the energy absorbed by a round-wire torsion spring. Express the answer in terms of the volume of metal in the coils.

309. A carbon-steel (ASTM A230) torsion spring is to resist a force of 55 lb. at a radius of 2 in.; the mean diameter is to be $2\frac{1}{2}$ in. Compute (a) the diameter of the wire for average service, (b) the number of coils for a deflection of 180° under the given torque, (c) the energy the spring has absorbed when the force is 55 lb.

310. The same as **309**, except that the size of square wire is to be found.

311. The same as **309**, except that the wire is rectangular. Let the long dimension of a section be perpendicular to the axis of the spring and equal to twice the short dimension.

312. A pivoted roller follower is held in contact with the cam by a torsion spring. The moment exerted by the spring varies from 20 lb-in. to 50 lb-in. as the follower oscillates through 30°. The spring is made of AISI 6152 steel, OQT 1000°F. What should be the value of D_w, D_m, and N_c if the factor of safety is 1.75 based on the Soderberg line? Would this be a conservative or risky approach?

FLAT AND LEAF SPRINGS

313. Derive an equation for the energy absorbed by a cantilever spring of uniform strength as shown in Fig. 6.20, *Text*. Express the answer in terms of metal volume and compare it with the energy absorbed by a beam of constant width.

314. The same as **313**, except that the spring is a simple beam, Fig. 6.21, *Text*.

315. A cantilever flat spring of uniform strength, Fig. 6.20, *Text*, is to absorb an energy impact of 500 ft-lb. Let the thickness of the steel, AISI 1095, OQT 900°F, be $\frac{1}{2}$ in. and let the maximum stress be half of the yield strength. (a) Find the width b of the spring at the widest point in terms of the length L. Determine values of b for lengths of 36 in., 48 in., 60 in., and 72 in. (b) Determine the deflection of the spring for each set of values found in (a).

316. The same as **315**, except that the spring is a simple beam of uniform strength supported at the ends, Fig. 6.21, *Text*.

317. One of the carbon contacts on a circuit breaker is mounted on the free end of a phosphor-bronze beam ($\mu = 0.35$). This beam has the shape of the beam shown in Fig. 6.24, *Text*, with $b = 1$ in., $b' = \frac{9}{16}$ in., $L = 4\frac{1}{2}$ in., and $h = \frac{1}{16}$ in. When the contacts are closed, the beam deflects $\frac{3}{4}$ in. Compute (a) the force on the contacts, (b) the maximum stress.

318. A cantilever leaf spring 26 in. long is to support a load of 175 lb. The construction is similar to that shown in Fig. 6.22(a), *Text*. The leaves are to be 2 in. wide, $\frac{3}{16}$ in. thick; SAE 9255 steel, OQT 1000°F; 10^7 cycles (§ 6.26). (a) How many leaves should be used if the surfaces are left as rolled? (b) The same as (a) except that the leaves are machined and the surfaces are not decarburized. (c) The same as (b), except that the surfaces are peened all over. (d) Which of these springs absorbs the most energy? Compute for each. (e) What are the load and deflection of the spring in (b) when the maximum stress is the standard-test yield strength?

319. The rear spring of an automobile has 9 leaves, each with an average thickness of 0.242 in. and a width of 2 in.; material is SAE 9261, OQT 1000°F. The length of the spring is 56 in. and the total weight on the spring is 1300 lb. Assume the spring to have the form shown in Fig. 6.22(b), *Text*. Determine (a) the rate of the spring, (b) the maximum stress caused by the dead weight. (c) What approxi-mate repeated maximum force (0 to F_{max}) would cause impending fatigue in 10^5 cycles, the number of applications of the maximum load expected during the ordinary life of a car? (If the leaves are cold rolled to induce a residual compressive stress on the surfaces, the endurance *limit* as $s_u/2$ should be conservative.)

320. The same as **319**, except that the spring is a cantilever type, Fig. 6.22(a), with a length of 25 in. and a static load of 650 lb.

321. The front spring of an automobile is made similar to Fig. 6.23, *Text*. Its length is 36 in.; width, 2 in., the average thickness for each of the 6 leaves, 0.213 in.; material is SAE 9255, OQT 1000°F. The load caused by the weight of the car is 775 lb. (a) What stress is caused by a force twice the dead weight? (b) What load would stress the spring to the yield strength? (c) Same as in **319**.

322. The same as **321**, except that the spring is cantilever type, Fig. 6.25, with a length of 16 in. and the weight load is 400 lb.

323–330. These numbers may be used for other problems.

Section 5

COLUMNS

MISCELLANEOUS

331. Determine the derivative $d(F_c/A)/d(L_e/k)$ from the Euler equation and equate it to the corresponding derivative from the Johnson equation. Show that the resulting L_e/k is the same as that in equation (e), § 7.6, *Text*. Show that for this value of L_e/k, the value of F_c/A is $s_y/2$ from both equations. Find the value of L_e/k for which both equations give the same critical load for (b) structural steel, AISI C1020, as rolled; (c) nickel-chromium steel, AISI 3140, OQT 900°F.

332. For the link of Fig. 7.3, p. 215, *Text*, let $L_e = L/1.41$ in the plane (b) and $L_e = L$ in the plane (a). What should be the relation between the width w and the thickness t of the link for the same resistance to buckling in each plane?

333. What value of L_e/k would result in the same safe load, $N = 3$, from the Euler equation as that obtained from the straight-line formula (c) in § 7.5, *Text?*

DESIGN PROBLEMS

334. A round steel rod made of structural steel, AISI C1020, as rolled, is to be used as a column, centrally loaded with 10 kips; $N = 3$. Determine the diameter for (a) $L = 25$ in., (b) $L = 50$ in. (c) The same as (a) and (b) except that the material is AISI 8640, OQT 1000°F. Is there any advantage in using this material rather than structural steel?

335. A hollow circular column, made of AISI C1020 structural steel, as rolled, is to support a load of 10,000 lb. Let $L = 40$ in., $D_i = 0.75D_o$, and $N = 3$. Determine D_o by (a) using either Euler's or the parabolic equation; (b) using the straight-line equation. (c) What factor of safety is given by the secant formula for the dimensions found in (a)?

336. A column is to be built up of $\frac{1}{2}$-in., AISI C1020, rolled-steel plates, into a square box-section. It is 6 ft. long and centrally loaded to 80,000 lb. (a) Determine the size of section for $N = 2.74$. (b) Compute N from the secant formula for the size found and compare with 2.74.

337. A column is to be made of $\frac{1}{2}$-in. structural steel plates (AISI 1020, as rolled), welded into an I-section as shown in Table AT 1 with $G = H$. The column, 15 ft. long, is to support a load of 125 kips. (a) Determine the cross-sectional dimensions from the straight-line equation. (b) Using either Johnson's or Euler's equation, compute the equiva-

33

lent stress and the factor of safety. (c) Compute N from the secant formula.

338. The link shown is to be designed for $N = 2.5$ to support an axial compressive load that varies from 0 to 15 kips; $L = 20$ in.; Material AISI 1030 steel, as rolled. (a) Determine the diameter considering buckling only. (b) Determine the diameter considering varying stresses and using the Soderberg line (perhaps too conservative). Estimate an appropriate strength-reduction factor (see Fig. AF 6). (c) Keeping in mind that the stress is always compressive, do you think that the answer from (a) will do? Discuss.

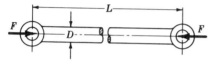

Problem 338.

339. The connecting link for a machine (see figure) is subjected to a load that varies from $+450$ (tension) to -350 lb. The cross section is to have the proportions $G = 0.4H, t = 0.1H$, fillet radius $r \approx 0.05H$; $L = 10$ in.; material, AISI C1020, as rolled. (a) Considering buckling only, determine the dimensions for a design factor of 2.5. (b) For the dimensions found compute the factor of safety from the Soderberg criterion.

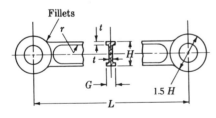

Problems 339, 340.

340. The same as **339**, except that the cross section is to be a solid rectangle of width $G = 0.4H$.

CHECK PROBLEMS

341. The link shown is subjected to an axial compressive load of 15 kips. Made of AISI C1030, as rolled, it has sec-

tional dimensions of $\frac{3}{4} \times 1\frac{3}{4}$ in. and a length of 20 in. Assume a loose fit with the pins. What is (a) the critical load for this column, (b) the design factor, (c) the equivalent stress, (d) the actual average stress under a load of 15 kips? What material does the secant formula indicate as satisfactory for the foregoing critical load when (e) $ec/k^2 = 0.25$, (f) $e = L_e/400$.

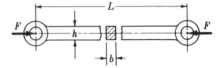

Problems 341, 342.

342. The same as **341** except that $L = 30$ in.

343. A schedule-40, 4-in. pipe is used as a column. Some of its properties are: $D_o = 4.5$ in., $D_i = 4.026$ in., $I = 7.233$ in.⁴, $k = 1.509$ in., $A = 3.174$ sq. in., $L = 15$ ft.; material equivalent to AISI C1015, as rolled. The total load to be carried is 200 kips. (a) What minimum number of these columns should be used if a design factor of 2.5 is desired and the load is evenly distributed among them? For the approximately fixed ends, use $L_e = 0.65L$ as recommended by AISC. (b) What is the equivalent stress in the column?

344. A centrally loaded column is a 10-in. × 49-lb., wide-flange I-beam whose properties are (see figure): $k_x = 4.35$ in., $k_y = 2.54$ in., area $A = 14.4$ sq. in., $I_x = 272.9$ in.⁴, $I_y = 93.0$ in.⁴; length $L = 30$ ft.; material AISI 1022, as rolled. Let the ends be a "little" fixed with $L_e = 0.8L$ and determine the critical load (a) according to the Johnson or the Euler equation; (b) according to the secant formula if ec/k^2 is assumed to be 0.25.

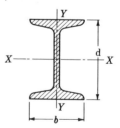

Problems 344-347.

345. The same as **344,** except that the material is aluminum alloy 2040-T4.

346. The same as **344,** except that the material is magnesium alloy AZ 91C, and equation (j), § 7.12, is also used in part (a).

347. The same as **344** except that the length is 14 ft.

348. A 4 $\times$ 3 $\times$ ½-in. angle is used as a flat-ended column, 5 ft. long, with the resultant load passing through the centroid G (see figure); $k_x = 1.25$ in., $k_y = 0.86$ in., $k_u = 1.37$ in., $k_v = 0.64$ in., $A = 3.25$ sq. in. Find the safe load if $N = 2.8$ and the material is (a) structural steel, (b) magnesium alloy AZ 91C (§ 7.12, *Text*), (c) magnesium alloy AZ 80 A, (d) magnesium alloy AZ 80 A as before, but use the Johnson formula and compare.

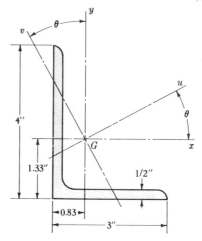

Problem 348.

349–360. These numbers may be used for other problems.

Section 6

COMBINED STRESSES

ECCENTRIC LOADING (NORMAL STRESSES)

DESIGN PROBLEMS

361. It is necessary to shape a certain link as shown in order to prevent interference with another part of the machine. It is to support a steady tensile load of 2500 lb. with a design factor of 2 based on the yield strength. The bottom edge of the midsection is displaced upward a distance $a = 2\frac{1}{2}$ in. above the line of action of the load. For AISI C1022, as rolled, and $h \approx 3b$, what should be h and b?

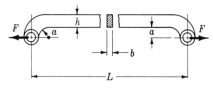

Problems 361-363.

362. A tensile load on a link as described in **361** varies from 0 to 3000 lb.; it is machined from AISI 1045, as rolled, and the lower edge of the link is $a = 0.5$ in. above the center line of the pins; $h \approx 3b$. Determine the dimensions of the link for $N = 2$ based on the Soderberg line.

36

363. The same as **362**, except that the load continuously reverses, 3 kips to -3 kips.

364. A circular column (see Fig. 8.3, *Text*), the material of which is SAE 1020, as rolled, is to have a length of 9 ft. and support an eccentric load of 16 kips at a distance of 3 in. from center line. Let $N = 3$. (a) What should be the outside diameter D_o if the column is hollow and $D_i = 0.75D_o$? (b) What should be the diameter if the column is solid?

365. The same as **364**, except that the length is 15 ft.

366. A link similar to the one shown is to be designed for: steady load $F = 8$ kips, $L = 20$ in. $\theta = 30°$; aluminum

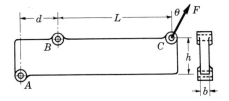

Problems 366, 372.

alloy 2024-T4; $N = 2.6$ on the yield strength. It seems desirable for the dimension b not to exceed $1\frac{3}{8}$ in. Determine b and h and check their proportions for reasonableness. The support is made so that the pin at B carries the entire horizontal component of F.

367. A column 15 ft. long is to support a load $F_2 = 50,000$ lb. acting at a distance of $e = 8$ in. from the axis of the column as shown (with $F_1 = 0$). Select a suitable I-beam for a design factor of 3 based on yield strength. The upper end of the column is free. See handbook for the properties of rolled sections.

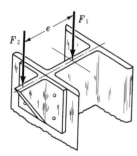

Problems 367, 368, 377.

368. The same as **367**, except that $F_1 = 50,000$ lb.

CHECK PROBLEMS

369. A cam press, similar to that of Fig. 19.1, *Text*, exerts a force of 10 kips at a distance of 7 in. from the inside edge of the plates that make up the frame. If these plates are 1 in. thick and the horizontal section has a depth of 6 in., what will be the maximum stress in this section?

370. A manufacturer decides to market a line of aluminum alloy (6061-T6) C-clamps, (see Fig. 8.4, *Text*). One frame has a T-section with the following dimensions (letters as in Table AT 1): $H = 1\frac{1}{16}$, $B = 1\frac{7}{32}$, $a = \frac{1}{8}$, and $t = \frac{1}{8}$ in. The center line of the screw is $2\frac{3}{8}$ in. from the inside face of the frame. (a) For $N = 3$ on the yield strength, what is the capacity of the clamp (gripping force)? (b) Above what approximate load will a permanent deformation of the clamp occur?

371. A C-frame (Fig. 8.5 *Text*) of a hand-screw press is made of annealed cast steel, ASTM A27-58 and has a section similar to that shown. The force F acts normal to the plane of the section at a distance of 12 in. from the inside face. The various dimensions of the sections are: $a = 3$ in., $b = 6$ in., $h = 5$ in., $d = e = f = 1$ in. Determine the force F for $N = 6$ based on the ultimate strength.

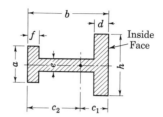

Problem 371.

72. In the link shown (**366**), let $b = \frac{1}{2}$ in., $h = 2$ in., $d = 2$ in., $L = 18$ in., and $\theta = 60°$. The clearance at the pins A and B are such that B resists the entire horizontal component of F; material is AISI C1020, as rolled. What may be the value of F for $N = 3$ based on the yield strength?

373. The link shown is subjected to a steady load $F_1 = 2.1$ kips; $b = 0.5$ in., $h = a = d = 2$ in., $L = 18$ in.; material AISI 1040, cold drawn (10% work). The dimensions are such that all of the horizontal reaction from F_2 occurs at A; and F_2 varies from 0 to a maximum, acting towards the right. For $N = 1.5$ based on the Soderberg line, what is the maximum value of F_2? Assume that the stress concentration at the holes can be neglected.

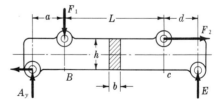

Problems 373-375.

374. The same as **373**, except that $F_1 = 3500$ lb.

375. The same as **373,** except that F_2 always acts horizontally toward the left varying from zero to a maximum.

376. A free-end column as shown, $L = 12$ ft. long, is made of 10-in. pipe, schedule 40, ($D_o = 10.75$ in., $D_i = 10.02$ in., $k = 3.67$ in., $A_m = 11.908$ in.^{2}, $I = 160.7$ in.^{4}, $Z = 29.9$ in.^{3}). The load completely reverses and $e = 15$ in.; $N = 3$; material is similar to AISI C1015, as rolled. (a) Using the equivalent-stress approach, compute the safe (static) load as a column only. (b) Judging the varying loading by the Soderberg criterion, compute the safe maximum load. (c) Determine the safe load from the secant formula. (d) Specify what you consider to be a reasonable safe loading.

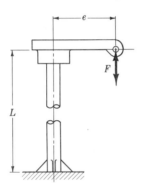

Problem 376.

377. A bracket is attached as shown (**367**) onto a 14-in. × 193-lb., wide-flange I-beam ($A = 56.73$ sq. in., depth = 15.5 in., flange width = 15.710 in., $I_{max} = 2402.4$ in.^{4}, $I_{min} = 930.1$ in.^{4}, $k_{min} = 4.05$ in.). The member is an eccentrically loaded column, 40 ft. long, with no central load ($F_1 = 0$) and no restraint at the top. For $e = 12$ in. and $N = 4$, what may be the value of F_2?

378. A 14-in. × 193-lb., wide-flange I-beam is used as a column with one end free ($A = 56.73$ sq. in., depth = 15.5 in., $I_{max} = 2402.4$ in.^{4}, $I_{min} = 930.1$ in.^{4}, $k_{min} = 4.05$ in., length $L = 40$ ft.). If a load F_2 is supported as shown on a bracket at an eccentricity $e = 4$ in. (with $F_1 = 0$), what may be its value for a design factor of 4? Flange width = 15.71 in.

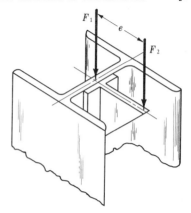

Problems 378, 379.

379. The same as **378,** except that $F_1 = 0.5F_2$.

380. The cast-steel link (SAE 080) shown (solid lines) is subjected to a steady axial tensile load and was originally made with a rectangular cross section, $h = 2$ in., $b = \frac{1}{2}$ in., but was found to be too weak. Someone decided to strengthen it by using a T-section (dotted addition), with h and b as given above. (a) Will this change increase the strength? Explain. (b) What tensile load could each link carry with $N = 3$ based on yield?

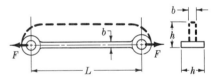

Problem 380.

COPLANAR SHEAR STRESSES

381. The figure shows a plate riveted to a vertical surface by 5 rivets.

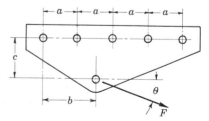

Problems 381, 382, 385, 386.

The material of the plate and rivets is SAE 1020, as rolled. The load $F = 5000$ lb., $b = 3$ in., $\theta = 0$, and $c = 5$ in.; let $a = 3D$. Determine the diameter D of the rivets and the thickness of plate for a design factor of 3 based on yield strengths.

382. The same as **381,** except that $\theta = 30°$.

383. Design a riveted connection, similar to that shown, to support a steady vertical load of $F = 1500$ lb. when $L = 18$ in. and $\theta = 0°$. Let the maximum spacing of the rivets, horizontally and vertically, be $6D$, where D is the diameter of the rivet; SAE 1020, as rolled, is used for all parts; $N = 2.5$ based on yield. The assembly will be such that there is virtually no twisting of the channel. The dimensions to determine at this time are: rivet diameter and minimum thickness of the plate.

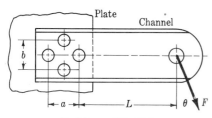

Problems 383, 384.

384. The same as **383,** except that $\theta = 45°$.

385. The plate shown **(381)** is made of SAE 1020 steel, as rolled, and held in place by five $\frac{3}{4}$ in. rivets that are made of SAE 1022 steel, as rolled. The thickness of the plate is $\frac{1}{2}$ in., $a = 2\frac{1}{2}$ in., $c = 6$ in., $b = 4$ in., and $\theta = 0$. Find the value of F for a design factor of 5 based on the ultimate strength.

386. The same as **385,** except that $\theta = 90°$.

387. The plate shown is made of AISI 1020 steel, as rolled, and is fastened to an I-beam (AISI 1020, as rolled) by three rivets that are made of a steel equivalent to AISI C1015, cold drawn. The thickness of the plate and of the flanges of the I-beam is $\frac{1}{2}$ in., the diameter of the rivets is $\frac{3}{4}$ in., $a = 8.5$ in.,

$b = 11.5$ in. and $c = 4.5$ in., $d = 4$ in. For $F_2 = 0$, calculate the value of F_1 for $N = 2.5$ based on yield strength.

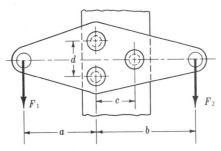

Problems 387, 388.

388. The same as **387,** except that $F_1 = 0$, and the value of F_2 is calculated.

NORMAL STRESSES WITH SHEAR

DESIGN PROBLEMS

389. The bracket shown is held in place by three bolts as shown. Let $a = 5\frac{1}{4}$ in., $\theta = 30°$, $F = 1500$ lb.; bolt material is equivalent to C1022, as rolled. (a) Compute the size of the bolts by equation (5.1), *Text.* (b) Assuming that the connecting parts are virtually rigid and that the initial stress in the bolts is about $0.7s_y$, compute the factor of safety by (i) the maximum shear stress theory, (ii) the octahedral shear theory. (c) Compute the maximum normal stress.

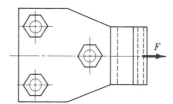

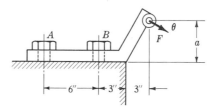

Problem 389.

390. For the mounted bracket shown, determine the rivet diameter (all same size) for $N = 3$, the design being for the external loading (initial stress ignored); $F = 2.3$ kips, $\theta = 0$, $c = 17$ in., $a = 1\frac{1}{2}$ in., $b = 14\frac{1}{2}$ in.; rivet material is AISI 1015, as rolled. Compute for (a) the maximum shear theory, (b) the maximum normal stress theory, (c) the octahedral shear theory.

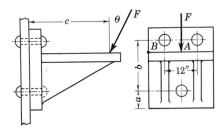

Problems 390-393, 396, 397.

391. The same as **390**, except that the angle $\theta = 30°$.

392. The same as **390**, except that the two top rivets are 2 in. long and the bottom rivet is $1\frac{1}{4}$ in. long.

393. The same as **390**, except that the load is applied vertically at B instead of at A; let $AB = 8$ in. The two top rivets are 12 in. apart.

394. The bracket shown is made of SAE 1020, as rolled, and the rivets are SAE 1015, cold drawn. The force $F = 20$ kips, $L = 7$ in., and $\theta = 60°$. Let the design factor (on yield) be 2. (a) Determine the thickness t of the arm. (b) Compute the rivet diameter by both maximum shear and octahedral shear theories and specify a standard size. (c) Decide upon a proper spacing of rivets and sketch the bracket approximately to scale. Is some adjustment of dimensions

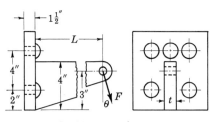

Problems 394, 395.

desirable? Give suggestions, if any. (No additional calculations unless your instructor asks for a complete design.)

395. The same as **394**, except that $\theta = 0$.

CHECK PROBLEMS

396. (a) If the rivets supporting the bracket of **390** are $\frac{5}{8}$ in. in diameter, $\theta = 0$, $c = 14$ in., $a = 2$ in., and $b = 18$ in., what are the maximum tensile and shear stresses in the rivets induced by a load of $F = 10$ kips. (b) For rivets of naval brass, $\frac{1}{4}$ hard, compute the factor of safety by maximum shear and octahedral shear theories (initial tension ignored).

397. The same as **396**, except that the two top rivets are $\frac{3}{4}$ in. in diameter and the bottom one is $\frac{1}{2}$ in. in diameter.

398. What static load F may be supported by the $\frac{3}{4}$-in. rivets shown, made of cold-finished C1015, with $N = 3$; $\theta = 0$, $a = 1\frac{1}{2}$, $b = 9$, $c = 14$, $f = 9$, $g = 12$ in.? Count on no help from friction and ignore the initial tension. Check by both maximum shear and octahedral shear theories.

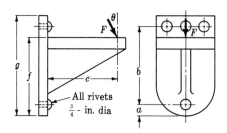

Problem 398.

399. The 2-in., UNC cap screw shown has been subjected to a tightening torque of 20 in-kips. The force $F = 12$ kips, $\theta = 60°$, and $Q = 0$; $L = 24$ in., $a = 20$ in., $b = 15$ in.; screw material is AISI C1137, as rolled. (a) What is the approximate initial tightening load? (b) What is the increase in this load caused by the external force F if the bar is 8 in. wide and 2 in. thick and the unthreaded shank of the screw is 2 in. long? (See § 5.9, *Text.*) (c) What are the maximum tensile and shear stresses in the bolt? (d) Compute the factor of safety from maximum normal stress, maximum shear, and octahedral shear theories.

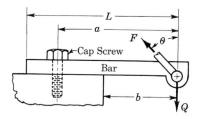

Problems 399, 421.

400. The plate shown is attached by three ½-in., UNC cap screws that are made of ASTM A325, heat-treated bolt material; $L = 26$ in., $a = 6$ in., $b = 4$ in., $\theta = 0$. The shear on the screws is across the threads and they have been tightened to an initial tension of $0.6 s_p$ (s_p = proof stress, § 5.8, *Text*). Which screw is subjected to (a) the largest force, (b) the largest stress? What safe static load can be supported by the screws for $N = 1.5$ based on the Hencky-Mises criterion?

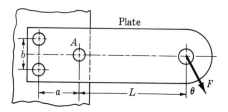

Problems 400, 401.

401. The same as **400**, except that the cap screw A is ¾ in. in diameter.

NORMAL STRESSES WITH TORSION

DESIGN PROBLEMS

402. A section of a machined shaft is subjected to a maximum bending moment of 70,000 in-lb., a torque of 50,000 in-lb., and an end thrust of 25,000 lb. The unsupported length is 3 ft. and the material is AISI C1030, normalized. Since the computations are to be as though the stresses were steady, use $N = 3.3$. Compute the diameter from both the maximum-shear and the octahedral-shear theories and specify a standard size.

403. The same as **402**, except that the unsupported length is 15 ft. Do not overlook the moment due to the weight of the shaft, which acts in the same sense as the given bending moment.

404. A shaft is to be made in two sections, I and II, of diameters D_1 and D_2, somewhat as shown, machined from AISI 1045, annealed. It is expected that $a = 8$ in., $b = 24$ in., $L = 20$ in., and the load $Q = 2$ kips, so seldom repeated that the design is for a steady load. The factor of safety is to be 2.2 on the basis of the octahedral-shear theory and closely the same in each section. The ends A and B are restrained from twisting, but they are designed to support the balancing reactions from Q without other moments. Decide upon standard size for D_1 and D_2.

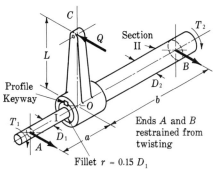

Fillet $r \approx 0.15 D_1$

Problems 404, 422.

405. The shaft shown overhangs a bearing on the right and has the following dimensions: $a = 5$ in., $b = \frac{1}{2}$ in., and $e = 10$ in. The material is AISI C1040, annealed. This shaft is subjected to a torque $T = 10,000$ in-lb., forces $F_1 = 10,000$ lb. and $F_2 = 20,000$ lb. Using a static-design approach, determine the diameter D for $N = 2.5$, with computa-

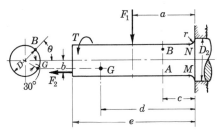

Problems 405-410.

tions from the maximum-shear and octahedral-shear theories.

406. The same as **405**, except that $F_2 = 0$.

CHECK PROBLEMS

407. The shaft shown overhangs a bearing at the right and has the following dimensions: $D = 2$ in., $a = 4$ in., $b = \frac{3}{4}$ in., $c = 2$ in., $d = 6$ in., $e = 8$ in., $r = \frac{1}{4}$ in. This shaft is subjected to a torque $T = 8000$ in-lb. and forces $F_1 = 8000$ lb., and $F_2 = 16,000$ lb. Determine the maximum-shear and normal stresses, and the octahedral-shear stress: (a) at points A and B ($\theta = 45°$), (b) at points M and N, (c) at point G.

408. The same as **407**, except that $F_2 = 0$.

409. The same as **407**, except that $T = 3\pi$ in-kips, $F_1 = 2\pi$ kips, and $F_2 = 5\pi$ kips.

410. The same as **407**, except that F_2 acts toward the right.

411. A 4-in. shaft carries an axial thrust of 20 kips. The maximum bending moment is $\frac{2}{3}$ of the twisting moment; material is AISI 8630, WQT 1100°F, and $N = 3$. Use the steady stress approach and compute the horsepower that may be transmitted at 2000 rpm?

412. The same as **411**, except that the shaft is hollow with an inside diameter of $2\frac{1}{2}$ in.

413. A hollow, alloy-steel shaft, AISI 4130, OQT 1100°F, has an OD of $3\frac{1}{4}$ in. and an ID of $2\frac{1}{2}$ in. It is transmitting 1500 hp at 1200 rpm, and at the same time is withstanding a maximum bending moment of 40,000 in-lb. and an axial compressive force $F = 10$ kips. The length of the shaft between bearings is 10 ft. Using a steady stress approach, determine (a) the maximum shearing stress in this shaft, (b) the maximum normal stress, (c) the factor of safety in each case. (d) Also compute N from the octahedral-shear theory.

VARYING STRESSES COMBINED

DESIGN PROBLEMS

414. The force F on the lever in the illustration (in the plane of the lever)

varies from a maximum of 424.2 lb. to a minimum of -141.4 lb.; $L = 20$ in., $a = 15$ in., $D_2 = 1.2D_1$, $r = 0.125D_1$, $\theta = 45°$; the material is cold-drawn SAE 1040, 10% worked, the design factor $N = 1.5$. Compute the diameter D_1 using the Soderberg-line approach with both the maximum-shear and octahedral-shear theories; indefinite life.

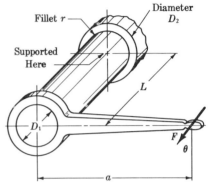

Problems 414-416.

415. The same as **414**, except that $F_{max} = 100$ lb., $F_{min} = 30$ lb., $\theta = 0$.

416. The same as **414**, except that $F_{max} = 300$ lb., $F_{min} = -100$ lb., $\theta = 0$.

417. A hollow steel shaft, SAE 1045, as rolled, has an inside diameter of one half of the outside diameter and is transmitting 1600 hp at 600 rpm. The maximum bending moment is 40,000 in-lb. Determine the diameter for $N = 3$ by both the maximum-shear and octahedral-shear theories. Specify a standard size. Use the Soderberg line for obtaining the equivalent stresses.

418. A section of a shaft without a keyway is subjected to a bending moment that varies sinusoidally from 30 to 15 then to 30 in-kips during two revolutions, and to a torque that varies similarly and in phase from 25 to 15 to 25 in-kips; there is also a constant axial force of 40 kips; the material is AISI 2340, OQT 1000°F; $N = 1.5$. Determine the diameter by (a) the maximum-shear-stress theory; (b) the octahedral-shear-stress theory.

419. The same as **418**, except that the shaft has a profile keyway at the point of maximum moment.

CHECK PROBLEMS

420. A 2-in. shaft made from AISI 1144, elevated temperature drawn, transmits 200 hp at 600 rpm. In addition to the data on the figure, the reactions are $B = 4.62$ kips and $E = 1.68$ kips. Compute the factor of safety by the maximum-shear and octahedral-shear theories.

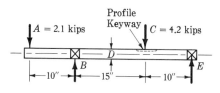

Profile Keyway

$A = 2.1$ kips $C = 4.2$ kips

B D E

← 10″ →← 15″ →← 10″ →

Problem 420.

421. In the figure **(399)**, the bar supports a static load $Q = 3000$ lb. acting down; $L = 16$ in., $a = 12$ in., $b = 7$ in. The force $F = 2500$ lb. is produced by a rotating unbalanced weight and is therefore repeated and reversed in both the horizontal and the vertical directions. The 1-in. cap screw, with cut UNC threads, is made of AISI C1137, annealed, and it has been subjected to a tightening torque of 4600 in-lb. The thickness of the bar is 2 in. (a) Compute the factor of safety for the load reversing in the vertical direction, and (b) in the horizontal direction (maximum-shear theory), with the conservative assumption that friction offers no resistance.

422. The load Q, as seen **(404)**, acts on the arm C and varies from 0 to 3 kips. The ends A and B of the shaft are restrained from turning through an angle but are supported to take the reactions A and B without other moments. The shaft is machined from AISI 1045, as rolled; $D_1 = 2$, $D_2 = 2.5$, $L = 15$, $a = 10$, $b = 20$ in. For calculation purposes, assume that the shaft size changes at the section of application of Q. Determine the factor of safety in accordance with the maximum-shear and octahedral-shear theories. Investigate both sections I and II. Would you judge the design to be 100% reliable?

423. A rotating shaft overhangs a bearing, as seen in the illustration. A

¼-in. hole is drilled at AB. The horizontal force F_2 varies in phase with the shaft rotation from 0 to 5 kips, but its line of action does not move. A steady torque $T = 8$ in-kips is applied at the end of the shaft; $D = 2$, $D_2 = 2.5$, $a = 2$, $b = 5$, $e = 0.5$, $r = \frac{1}{4}$ in. The material is AISI C1040, annealed. What steady vertical load F_1 can be added as shown if the design factor is to be 2.5 from the octahedral-shear theory? Assume that the cycling of F_2 may be such that the worst stress condition occurs at the hole.

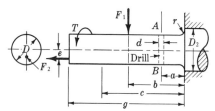

F_1

T A r

D d D_2

e Drill

F_2 B ←a→

←b→

←c→

←g→

Problem 423.

POWER SCREWS

424. Design a square-thread screw for a screw jack, similar to that shown, which is to raise and support a load of 5 tons. The maximum lift is to be 18 in. The material is AISI C1035, as rolled, and $N \approx 3.3$ based on the yield strength.

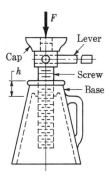

F

Cap

Lever

h

Screw

Base

Problems 424–426.

425. (a) For the screw of **424**, what length of threads h will be needed for a bearing pressure of 1800 psi? (b) Complete the design of the jack. Let the base be cast iron and the threads integral with the base. Devise a method of turning the screw with a round steel rod as a lever and fix the details of a nonrotating cap on which the load rests. (c) What should be the diameter of the rod used to turn the screw? If a man exerts a pull of 150 lb. at the end, how long must the rod be?

426. A screw jack, with a 1¼-in. square thread, supports a load of 6000 lb. The material of the screw is AISI C1022, as rolled, and the coefficient of friction for the threads is about 0.15. The maximum extension of the screw from the base is 15 in. (a) Considering the ends of the screw restrained so that $L_e = L$, find the equivalent stress and the design factor. (b) If the load on the jack is such that it may sway, the screw probably acts as a column with one end free and the other fixed. What is the equivalent stress and the factor of safety in this instance? (c) What force must be exerted at the end of a 20-in. lever to raise the load? (d) Find the number of threads and the length h of the threaded portion in the cast-iron base for a pressure of 500 psi on the threads. (e) What torque is necessary to lower the load?

427. A square-thread screw, 2 in. in diameter, is used to exert a force of 24,000 lb. in a shaft-straightening press. The maximum unsupported length of the screw is 16 in. and the material is AISI C1040, annealed. (a) What is the equivalent compressive stress in the screw? Is this a satisfactory value? (b) What torque is necessary to turn the screw against the load for $f = 0.15$? (c) What is the efficiency of the screw? (d) What torque is necessary to lower the load?

428. (a) A jack with a 2-in., square-thread screw is supporting a load of 20 kips. A single thread is used and the coefficient of friction may be as low as 0.10 or as high as 0.15. Will this screw always be self-locking? What torque is necessary to raise the load? What torque is necessary to lower the load? (b) The same as (a) except that a double thread is used. (c) The same as (a) except that a triple thread is used.

429. The conditions for a self-locking screw are given in § 8.23, *Text*. Assume that the coefficient of friction is equal to the tangent of the lead angle and show that the efficiency of a self-locking screw is always less than 50%.

CURVED BEAMS

430. It is necessary to bend a certain link somewhat as shown in order to prevent interference with another part of the machine. It is estimated that sufficient clearance will be provided if the center line of the link is displaced $e = 3$ in. from the line of action of F, with a radius of curvature of $R \approx 5.5$ in., $L = 10$ in., material is wrought aluminum alloy 2014 T6; $N = 2$ on the basis of the maximum shear stress; $F = 2500$ lb. with the number of repetitions not exceeding 10^6. (a) If the section is round, what should be its diameter D? (b) If the link is bent to form cold, will the residual stresses be helpful or damaging? Discuss.

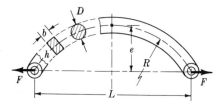

Problems 430, 431.

431. The same as **430**, except that the section is rectangular with $h \approx 3b$; see figure.

432. A hook is to be designed similar to that shown to support a maximum load $F = 2500$ lb. that will be repeated an indefinite number of times; the horizontal section is to be circular of radius c and the inside radius a is 1½ in. (a) Determine the diameter of the horizontal section for $N = 2$ based on the Soderberg line, if the material is AISI 4130, WQT 1100°F. (b) Calculate the value of the static load that produces incipient yielding.

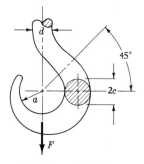

Problems 432-434.

433. The same as **432**, except that the hook is expected to be subjected to 100,000 repetitions of the maximum load.

434. A hook, similar to that shown with a horizontal circular section of diameter $2c$, is to be designed for a capacity of 2000 lb. maximum, a load that may be applied an indefinite number of times. A value of $a = 2$ in. should be satisfactory for the radius of curvature of the inside of the hook. Let $N = 1.8$ based on the modified Goodman line. At the outset of design, the engineer decided to try AISI C1040, OQT 1100°F. (a) Compute the diameter of the horizontal section. (b) If the 45° circular section is made the same diameter, what is its design factor (modified Goodman)? Could this section be made smaller or should it be larger?

435. A C-frame hand press is made of annealed cast steel (A27-58) and has a modified I-section, as shown. The dimensions of a 45° section CD are: $a = 3$, $b = 6$, $h = 4$, $t = 1$ in., radius $r = 1$ in.; also $g = 12$ in.; and the maximum force is $F = 17$ kips, repeated a relatively few times in the life of the press. (a) Applying the straight-beam formula to the 45° section, compute the maximum and minimum normal stresses. (b) Do the same, applying the curved-beam formula. (c) By what theory would you judge this section to have been designed? If the radius r were increased several times over, as it could have been done, would the stress have been materially reduced? Give reasons for your conclusions.

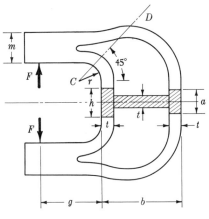

Problem 435.

436. A heavy C-clamp, similar to the figure, is made of normalized cast steel (A27-58) and has a T-section where $t = \frac{7}{16}$ in.; $q = 2\frac{3}{4}$, $a = 1\frac{3}{4}$ in. What is the safe capacity if $N = 2$ based on yield?

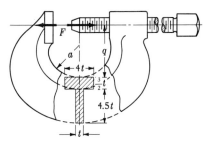

Problems 436, 437.

437. The same as **436**, except that the section is trapezoidal with $b = \frac{3}{4}$ in. (see figure). Ignore the effect of rounding off of the corners.

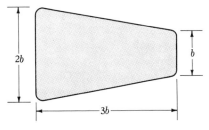

Problem 437.

THICK-SHELL CYLINDERS; INTERFERENCE FITS

438. Special welded steel pipe, equivalent in strength to SAE 1022, as rolled, is subjected to an internal pressure of 8000 psi. The internal diameter is to be $4\frac{1}{2}$ in. and the factor of safety is to be 3, including an allowance for the weld. (a) Find the thickness of the pipe according to the distortion-energy theory. (b) Using this thickness find the maximum normal and shear stresses and the corresponding safety factors. (c) Compute the thickness from the thin-shell formula and from the Barlow formula.

439. The internal diameter of the cast-steel cylinder, SAE 0030, of a hydraulic press is 12 in. The internal working pressure is 6000 psi, $N = 2.5$.

Find the thickness of the cylinder walls (a) from the maximum-shear-stress theory, (b) from the octahedral-shear theory. (c) Compute the thickness from the thin-shell and Barlow formulas. What do you recommend?

440. The same as **439**, except a higher-strength material is selected. Try cast-steel SAE 0105.

441. A $2\frac{1}{2}$ in. heavy-wall pipe has the following dimensions: $OD = 2.875$, $ID = 1.771$, $t = 0.552$ in.; inside surface area per foot of length = 66.82 in.^{2}, outside surface area per foot of length = 108.43 in.^{2}. The material is chromium-molybdenum alloy, for which the permissible tangential tensile stress is 15 ksi at temperatures between 700–800°F. (a) Compute the maximum internal working pressure for this pipe from Lame's formula, by the maximum-shear and octahedral-shear theories. (b) What is the stress at an external fiber? (c) A higher design stress would be permitted for an external pressure alone. Nevertheless, compute the external pressure corresponding to a maximum tangential stress of 15 ksi.

442. A cast-steel hub is to be shrunk on a 1.5-in., SAE 1035, as-rolled, steel shaft. The equivalent diameter of the hub is 2.5 in., its length is 4 in. (a) What must be the interference of metal if the holding power of this fit is equal to the torsional yield strength of the shaft? Use Baugher's recommendations. (b) What are the corresponding tangential and radial stresses in the hub?

443. The same as **442**, except that the hub is ASTM 20, cast iron. Will the resulting tensile stresses be safe for cast iron?

444. A cast-steel gear is pressed onto a 2-in. shaft made of AISI 3140, OQT 1000°F. The equivalent hub diameter is 4 in., and the hub length is 4 in. (a) What are the maximum tangential and radial stresses in the hub caused by a class FN 2 interference fit? Compute for the apparent maximum value of i (but recall the probability of this event). (b) What axial force F in tons will be required to press the gear on the shaft if f_1 is assumed to be 0.2? (c) What torque may the force fit safely transmit? (d) Is the holding capacity of this fit large enough to transmit a torque that produces a simple torsional stress of $0.6s_{ys}$ in the shaft?

445. The same as **442**, except that a class FN 4 fit is investigated and the computation is made for the average i.

446. A No. 217 ball bearing has a bore of 3.3465 in., a width of 1.1024 in., and the inner race is approximately $\frac{3}{8}$ in. thick. This bearing is to be mounted on a solid shaft with $i = 0.0014$. (a) Calculate the maximum radial and tangential stresses in the race. (b) Estimate the force required to press the bearing onto the shaft.

447. A steel disk of diameter D_o and thickness $L = 4$ in. is to be pressed onto a 2-in. steel shaft. The parts are manufactured with class FN 5 fit, but assembled parts are selected so as to give approximately the average interference. What will be the maximum radial and tangential stresses in the disk if (a) D_o is infinitely large; (b) $D_o = 10$ in.; (c) $D_o = 4$ in.; (d) $D_o = 2.5$ in.?

448. A steel cylinder is to have an inside diameter of 3 in. and $p_i = 30,000$ psi. (a) Calculate the tangential stresses at the inner and outer surfaces if the outside diameter is 6 in. (b) It was decided to make the cylinder in two parts, the inner cylinder with $D_1 = 3$ in. and $D_i = 4.5$ in., the outer cylinder with $D_i = 4.5$ in. and $D_o = 6$ in. (see figure).

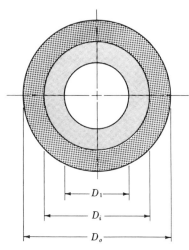

Problems 448, 449.

The two cylinders were shrunk together with $i = 0.003$ in. Calculate the pressure at the interface and the tangential stresses at the inner and outer surfaces of each cylinder. (*Suggestion:* first derive an equation for the interface pressure.)

449. A phosphor-bronze (B139C) bushing has an ID $= \frac{3}{4}$ in., an OD $= 1\frac{1}{4}$ in., and a length of 2 in. It is to be pressed into a cast-steel cylinder that has an outside diameter of $2\frac{1}{2}$ in. An ASA class FN 2 fit is to be used with selective assembly to give approximately the interference $i = 0.0016$ in. Calculate (a) p_i, (b) the maximum tangential stress in the steel cylinder, (c) the force required to press bushing into the cylinder, (d) the decrease of the inside diameter of the bushing.

DESIGN PROJECTS

450. A jib crane similar to the one shown is to be designed for a capacity of $F =$ ____ (say, 1 to 3 tons). The load F can be swung through 360°; $L \approx 10$ ft., $b \approx 8.5$ ft., $c \approx 2$ ft. The moment on the jib is balanced by a couple QQ on the post, the forces Q acting at supporting bearings. The crane will be fastened to the floor by 6 equally spaced bolts on a $D_1 = 30$-in. bolt circle; outside diameter of base $D_2 = 36$ in. (a) Choose a pipe size (handbooks) for the column such that the maximum equivalent stress does not exceed 12 ksi. (b) Choose an I-beam for the jib such that the maximum stress does not exceed 12 ksi. (c) Compute the maximum external load on a base bolt and decide upon the size. (d) Complete other detail as required by the instructor, such as: computing Q and choosing bearings (ball or roller?), the internal construction and assembly in this vicinity, detail sketches giving full information.

451. Design an air-operated punch press similar to the one shown. Let the

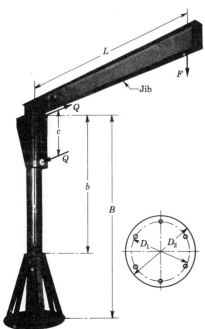

Courtesy Manning, Maxwell, and Moore, Inc.

Problem 450.

Courtesy Hannifin Mfg. Co., Chicago

Problem 451.

force at the punch be 12 tons, (or other capacity as specified by the instructor), the depth of throat to the inside edge of the frame be 25 in., the diameter and stroke of the piston about 8 in. by 8 in., the mechanical advantage of the lever about 7, and the diameter of the punch $\frac{3}{16}$ in. Determine first the horizontal section of the frame, and locate and design the cylinder. Then determine the relative arrangement of the various links and make a force analysis, from which the design of certain parts follow. Determine the actual distance of movement of the punch (not less than about 1 in.). The illustration will assist the student in settling upon the proportions of parts for which strength calculations cannot be made.

452. Design a screw press similar to that shown for a load of ___ (say, 3) tons on the screw. The depth of the throat g is to be ___ (10) in. and the height of the throat h is to be ___ (15) in. (The instructor will assign the data.) The order of procedure may be as follows: (a) Find the diameter of the screw. If $L_e/k > 40$, check as a column. If the top of the screw is squared off for a handwheel or handle, check this section for twisting. The equation for pivot friction, if desired, is in § 18.10, *Text.* (b) Decide upon the diameter of the handwheel or the length of handle (if one is needed to obtain the maximum pressure), assuming that the maximum force to be exerted by a man is about 150 lb. Dimensions of handwheels may be found in handbooks. The handle may be designed by the flexure formula. (c) Design the frame. The shape of the section of the frame will depend on the material used. A T-section is suitable for cast iron (say $N = 6$ on the ultimate strength), a hollow box or modified I-section is suitable for cast steel. The 45° section CD of the frame should be safe as a curved beam. See Table AT 18. In this connection, it will be well to make the radius r as large as practicable, since the larger r the less the stresses from a given load. Compute the dimensions of the vertical section. It is a good plan to keep t and a the same in all sections. (d) Design the bushing if one is used. The height b depends upon the number of threads in contact, which in turn depends upon the bearing pressure used in design. (Say half-hard yellow brass?) Compute

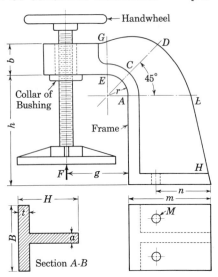

Problem 452.

the outside bushing diameter, the diameter and thickness of the collar, and decide upon dimensions to be used. (e) Fix the location and number of bolts to be used to fasten the frame to the base plate, and determine their size. Use a common bolt material. (f) Decide upon all other details of design. Make a separate sketch of each part of the machine showing thereon all dimensions necessary for manufacture. It is suggested that, first, all materials be tentatively decided upon, after which design stresses may be chosen. See that design stresses for the various parts bear a logical relation to one another. It is not necessary to follow this procedure in detail. It is likely that one will have to leave certain details unfinished from time to time, because these details depend on parts of the design not yet completed. *Make sure that all parts can be assembled after they are made.* Notice that the plate on the lower end of the screw must be connected in such a manner that the screw may turn while the plate does not.

453. Design a jib crane, as suggested by the illustrations, to lift a load W of ___ tons. The maximum radius of swing is to be about ___ ft. (The instructor will assign data). Suggested procedure: (a) From catalogues, select a hoist to suit the purpose, giving reasons for your choice, and noting significant

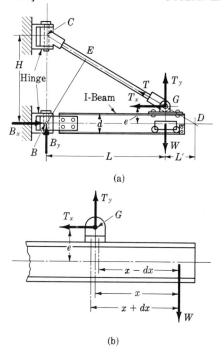

(a)

(b)

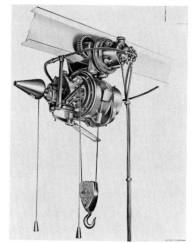

Courtesy Ingersoll-Rand Co.,
Philipsburg, N. J.

(c)

Problem 453.

dimensions. Of course, in the end, the hoist trolley has to match the size of I-beam used. (b) Let the angle that the diagonal tension rod makes with the horizontal be about 20° to 25°, and decide upon the dimensions H and L. Note that the point G does not necessarily have to be at the extreme position of the load. As a matter of fact, some advantage may result from having G inside the outermost position of the load. Make force analyses (including weight of hoist as part of load) for (1) the condition of maximum column action, (2) the condition of maximum bending moment on the beam, and (3) the condition for maximum force on the hinge B (to be used for the design of this hinge). (c) Find the size of I-beam such that the maximum stress for any position of the load falls within the limits of 12 and 15 ksi, usually by assuming a standard beam and checking the stress. According to the arrangement of parts, it may be necessary to design the connection at G between the rod and the beam first. With the details of this connection known and with the depth of beam assumed, the location of point G, the

point of application of the force T, can be determined. The bending moment of a section a minute distance to the right of G is $W(x - dx)$. A minute distance to the left of G, the bending moment is $W(x + dx) - T_x e - T_y\,dx$; that is, the moment changes suddenly at G by the amount $T_x e$. (d) Determine the size of diagonal support, including details of connections. (e) Design the connections at each end of the diagonal and the hinge at C. Settle upon the details including the method of attaching the hinge to the vertical surface, which may be a wide-flange beam. (f) Design the hinge at B and the connection to the I-beam; also the details of the method of attaching the hinge to the vertical surface. Where material is not specified, make your choice clear. There should be no doubt as to your design stresses or design factor. Show a neat large sketch, fully dimensioned, of each part separately. It is unlikely that too much detail will be shown.

454–470. These numbers may be used for other problems.

Section 7

SHAFT DESIGN

DESIGN PROBLEMS

471. A short stub shaft, made of SAE 1035, as rolled, receives 30 hp at 300 rpm via a 12-in. spur gear, the power being delivered to another shaft through a flexible coupling. The gear is keyed (profile keyway) midway between the bearings. The pressure angle of the gear teeth $\phi = 20°$; $N = 1.5$ based on the octahedral shear stress theory with varying stresses. (a) Neglecting the radial component R of the tooth load W, determine the shaft diameter. (b) Considering both the tangential and the radial components, compute the shaft diameter. (c) Is the difference in the results of the parts (a) and (b) enough to change your choice of the shaft size?

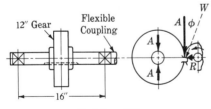

Problem 471.

472. A cold-finished shaft, AISI 1141, is to transmit power that varies from 200 to 100 and back to 200 hp in each revolution at a speed of 600 rpm. The

power is received by a 20-in. spur gear A and delivered by a 10-in. spur gear C. The tangential forces have each been converted into a force (A and C shown) and a couple (not shown). The radial component R of the tooth load is to be ignored in the initial design. Let $N = 2$ and, considering varying stresses with the maximum shear theory, compute the shaft diameter.

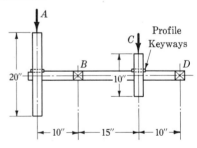

Problems 472-474.

473. The same as **472**, except that both gears have a pressure angle of $\phi = 20°$ and the radial (horizontal) forces are to be considered (see illustration for **471**).

474. The same as **472**, except that the force C acts upward (the mating gear is on the other side of C).

475. A shaft S, of cold-drawn AISI 1137, is to transmit power received from shaft W, which turns at 2000 rpm, through the 5-in. gear E and 15-in. gear A. The power is delivered by the 10-in. gear C to gear G, and it varies from 10 hp to 100 hp and back to 10 hp during each revolution of S. The design is to account for the varying stresses, with calculations based on the octahedral shear stress theory. Let $N = 1.8$ and compute the shaft diameter, using only the tangential driving loads for the first design.

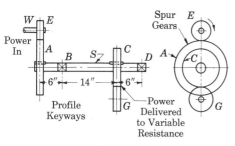

Problems 475-477.

476. The same as **475,** except that the radial components of the gear-tooth loads are to be accounted for; all gears have a pressure angle $\phi = 20°$ (see figure for **471**).

477. The same as **475,** except that gear G is above C.

478. A shaft made of AISI 1137, cold rolled, for a forage harvester is shown. Power is supplied to the shaft by a vertical flat belt on the pulley A. At B, the roller chain to the cutter exerts a force vertically upwards, and the V-belt to the blower at C exerts a force vertically upwards. At maximum operating conditions, the flat belt supplies 35 hp at 425 rpm, of which 25 hp is delivered to the cutter and 10 hp to the blower. The two sections of the shaft are joined by a flexible coupling at D and the various wheels are keyed (sled runner keyway) to the shafts. Allowing for the varying stresses on the basis of the von Mises-Hencky theory of failure, decide upon the diameters of the shafts. Choose a design factor that would include an allowance for rough loading.

479. A shaft for a punch press is supported by bearings D and E (with $L = 24$ in.) and receives 25 hp, while rotating at 250 rpm, from a flat-belt drive on a 44-in. pulley at B, the belt being at 45° with the vertical. An 8-in. gear at A delivers the power horizontally to the right for the punching operation. A 1500-lb. flywheel at C has a radius of gyration of 18 in. During punching, the shaft slows and energy for punching comes from the loss of kinetic energy of the flywheel in addition to the 25 hp constantly received via the belt. A reasonable assumption for design purposes would be that the power to A doubles during punching, 25 hp from the belt, 25 hp from the flywheel. The phase relations are such that a particular point in the section where the maximum moment occurs is subjected to alternating tension and compression. Sled runner keyways are used for A, B, and C; material is cold-drawn AISI 1137; use a design factor of $N = 2.5$ with the octahedral shear theory and account for the varying stresses. Determine the shaft diameter.

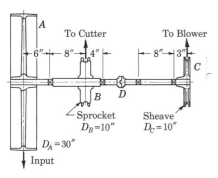

Problem 478.

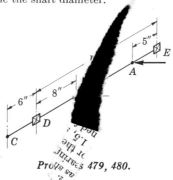

Probs. 479, 480.

480. The same as **479,** except that gear A is 5 in. to the *right* of bearing E.

THRUST LOADS

481. A cold-drawn monel propeller shaft for a launch is to transmit 400 hp at 1500 rpm without being subjected to a significant bending moment; and $L_e/k < 40$. The efficiency of the propeller is 70% at 30 knots (1.152 mph/knot). Consider that the number of repetitions of the maximum power at the given speed is 2×10^5. Let $N = 2$ based on the maximum shear theory with varying stresses. Compute the shaft diameter.

482. A shaft receives 300 hp, while rotating at 600 rpm, through a pair of bevel gears, and it delivers this power via a flexible coupling at the other end. The shaft is designed with the average forces (at the midpoint of the bevel-gear face); the tangential driving force is F, $G = 580$ lb., $Q = 926$ lb., which are the rectangular components of the total reaction between the teeth; $D_m = 24$ in., $L = 36$ in., $a = 10$ in. Let the material be AISI C1045, cold drawn; $N = 2$. Considering varying stresses and using the octahedral shear theory, determine the shaft diameter.

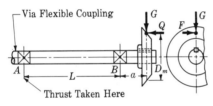

Problems 482, 485, 486.

483. The worm shown is to deliver 65.5 hp steadily at 1750 rpm. It will be integral with the shaft if the shaft size needed permits, and its pitch diameter 3 in. The 12-in. pulley receives the power from a horizontal belt in which the tight tension $F_1 = 2.$ The forces (in kips) on the worm are shown, with the axial force taken by bearing B. The strength reduction factor for thread roots may be taken as $K_f =$ shear or bending. The shaft is machined from AISI 1045, as rolled. (a) For $N = 2.2$ (Soderberg criterion) by the octahedral-shear theory, compute the required minimum diameter at the root of the worm thread (a first approximation). (b) What should be the diameter of the shaft 2.5 in. to the left of

the center line of the worm? (c) Select a shaft size D and check it at the pulley A.

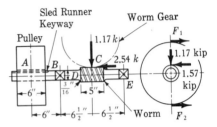

Problem 483.

484. A propeller shaft as shown is to receive 300 hp at 315 rpm from the right through a flexible coupling. A 16-in. pulley is used to drive an auxiliary, taking 25 hp. The belt pull F_B is vertically upward. The remainder of the power is delivered to a propeller that is expected to convert 60% of it into work driving the boat, at which time the boat speed is 1500 fpm. The thrust is to be taken by the right-hand bearing. Let $N = 2$; material, cold-worked stainless 410. Use the octahedral shear theory with varying stresses. (a) Determine the shaft size needed assuming no buckling. (b) Compute the equivalent column stress. Is this different enough to call for another shaft size? Compute N by the maximum shear stress theory, from both equations (8.4) and (8.10).

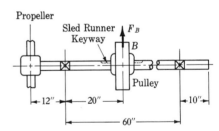

Problem 484.

CHECK PROBLEMS

485. A 3-in. rotating shaft somewhat as shown (**482**) carries a bevel gear whose mean diameter is $D_m = 10$ in. and which is keyed (profile) to the *left* end. Acting on the gear are a radial force $G = 1570.8$ lb., a driving force $F = 6283.2$ lb., and a thrust force $Q = 3141.6$

lb. The thrust force is taken by the right-hand bearing. Let $a = 5$ in. and $L = 15$ in.; material, AISI C1040, annealed. Base calculations on the maximum shearing stress theory with variable stresses. Compute the indicated design factor N. With the use of a sketch, indicate the exact *point* at which maximum normal stress occurs.

486. The same as **485** except that the material is AISI 4150, OQT 1000°F.

487. A $2\frac{7}{16}$-in. countershaft in a machine shop transmits 52 hp at 315 rpm. It is made of AISI 1117, as rolled, and supported upon bearings A and B, 59-in. apart. Pulley C receives the power via a horizontal belt, and pulley D delivers it vertically downward, as shown. Calculate N based on the octahedral-shear-stress theory considering varying stresses.

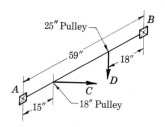

Problems 487, 488.

488. The same as **487** except that hp = 100 and the material is AISI 1144, elevated temperature drawn (ETD).

489. A shaft for a general-purpose gear-reduction unit supports two gears as shown. The 5.75-in. gear B receives 7 hp at 250 rpm. The 2.25-in. gear A delivers the power, with the forces on the shaft acting as shown; the gear teeth have a pressure angle of $\phi = 14\frac{1}{2}°$ (tan $\phi = A_h/A_v = B_h/B_v$). Both gears are keyed (profile) to the shaft of AISI 1141, cold

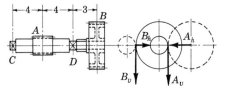

Problems 489, 490.

rolled. (a) If the fillet radius is $\frac{1}{8}$ in. at bearing D, where the diameter is $1\frac{3}{8}$ in., compute N based on the octahedral-shear-stress theory (Soderberg line). The shaft diameter at A is $1\frac{11}{16}$ in. What is N here?

490. The same as **489**, except that $\phi = 20°$.

THRUST LOADS

491. The high-speed shaft of a worm-gear speed reducer, made of carburized AISI 8620, SOQT 450°F, is subjected to a torque of 21,400 in-lb. applied to the right end with no bending. The force on the worm has three components: a horizontal force opposing rotation of $W = 6180$ lb., a vertical radial force $S = 1940$ lb., and a rightward thrust force of $F = 6580$ lb. The shaft has the following dimensions: $a = 6$, $b = 4\frac{7}{8}$, $c = 10$, $d = 4\frac{9}{16}, e = 2\frac{3}{4}, f = 13\frac{9}{16}, g = 11.646$, $h = 10.370$, $D_1 = 3.740$, $D_2 = 4\frac{13}{16}$, $D_3 = 4$, $D_4 = 3.3469$, $D_5 = 3.253$, $r_1 = 0.098, r_2 = r_3 = \frac{3}{4}, r_4 = 0.098, r_5 = \frac{1}{16}$, all in inches. The pitch diameter of the worm, 6.923 in., is the effective diameter for the point of application of the forces. The root diameter, 5.701 in., is used for stress calculations. The left-hand bearing takes the thrust load. Calculate N based on the octahedral-shear-stress theory with varying stresses. (*Data courtesy of Cleveland Worm & Gear Company.*)

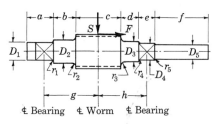

Problem 491.

492. The slow-speed shaft of a speed reducer shown, made of AISI 4140, OQT 1200°F, transmits 100 hp at a speed of 388 rpm. It receives power through a 13.6 in. gear B. The force on this gear has three components: a horizontal tangential driving force $F_t = 2390$ lb., a vertical radial force $S = 870$ lb., and a thrust force $Q = 598$ lb. taken by the right-hand bearing. The power is delivered to a belt

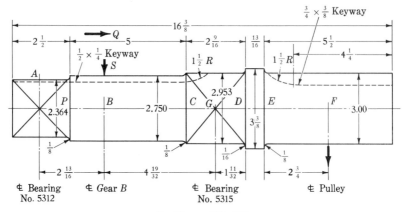

₵ Bearing   ₵ Gear B   ₵ Bearing   ₵ Pulley
No. 5312   No. 5315

Problems 492, 493.

at F that exerts a downward vertical force of 1620 lb.; sled runner keyways. Use the octahedral shear theory with the Soderburg line and compute N at sections C and D. (*Data courtesy of Twin Disc Clutch Company.*)

493. The same as **492**, except that the direction of the belt force is unknown, since it is to be determined by the way the customer uses the reducer. Assume it to be in the direction that will produce the maximum bending moment in the shaft.

TRANSVERSE DEFLECTIONS*

494. The forces on the 2-in. steel shaft shown are $A = 2$ kips, $C = 4$ kips. Determine the maximum deflection and the shaft's slope at D.

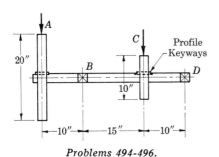

Problems 494-496.

* The instructor may have the student find the deflections, or shaft diameters for a permissible deflection, in any of the previous problems.

495. The forces on the steel shaft shown are $A = 2$ kips, $C = 4$ kips. Determine the constant shaft diameter that corresponds to a maximum deflection of 0.006 in. at section C.

496. The forces on the steel shaft shown are $A = 2$ kips, $C = 4$ kips. Determine a constant shaft diameter that would limit the maximum deflection at section A to 0.003 in.

497. A steel shaft is loaded as shown and supported in bearings at R_1 and R_2. Determine (a) the slopes at the bearings and (b) the maximum deflection.

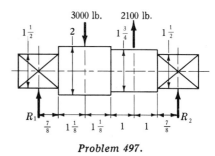

Problem 497.

498. (a) Determine the diameter of the steel shaft shown if the maximum deflection is to be 0.01 in.; $C = 1.5$ kips, vertically downward, $B = 0.6$ kips, $A = 1.58$ kips, $L = 24$ in. (b) What is the slope of the shaft at bearing D? See **479.**

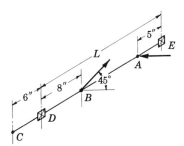

Problems 498, 505, 506.

CRITICAL SPEED

499. A small, high-speed steam turbine has a single disk, weighing 0.85 lb., mounted at the midpoint of a 0.178-in. shaft, whose length between bearings is $6\frac{1}{2}$ in. What is the critical speed if the shaft is considered as simply supported?

500. The bearings on a $1\frac{1}{2}$-in. shaft are 30 in. apart. On the shaft are three 300-lb. disks, symmetrically placed 7.5 in. apart. What is the critical speed of the shaft?

501. A fan for an air-conditioning unit has two 50-lb. rotors mounted on a 3-in. steel shaft, each being 22 in. from an end of the shaft which is 80 in. long and simply supported at the ends. Determine (a) the deflection curve of the shaft considering its weight as well as the weight of the rotors, (b) its critical speed.

ASME CODE

502. A cold-rolled transmission shaft, made of annealed AISI C1050, is to transmit a torque of 27 in-kips with a maximum bending moment of 43 in-kips. What should be the diameter according to the *Code* for a mild shock load?

503. A machinery shaft is to transmit 82 hp at a speed of 1150 rpm with mild shock. The shaft is subjected to a maximum bending moment of 7500 in-lb. and an axial thrust load of 15,000 lb. The material is AISI 3150, OQT 1000°F. (a) What should be its diameter when designed according to the *Code?* (b) Determine the corresponding conventional factor of safety (static-approach and maximum-shear theory).

504. A short stub shaft, made of SAE 1035, as rolled, receives 30 hp at 300 rpm via a 12-in. spur gear, the power being delivered to another shaft through a flexible coupling. The gear is keyed midway between the bearings and its pressure angle $\phi = 20°$. See the figure for **471.** (a) Neglecting the radial component of the tooth load, determine the shaft diameter for a mild shock load. (b) Considering both the tangential and radial components, compute the shaft diameter. (c) Is the difference in the foregoing results enough to change your choice of shaft size?

505. Two bearings D and E, a distance $L = 24$ in. apart, support a shaft for a punch press on which are an 8-in. gear A, a 44-in. pulley B, and a flywheel C, as indicated (**498**). Weight of flywheel is 1500 lb.; pulley B receives the power at an angle of 45° to the right of the vertical; gear A delivers it horizontally to the right. The maximum power of 25 hp at 250 rpm is delivered, with heavy shock. For cold-finish AISI 1137, find the diameter by the ASME Code.

506. The same as **505** except that the gear A is 5 in. to the right of the right-hand bearing.

507–520. These numbers may be used for other problems.

KEYS AND COUPLINGS

FLAT AND SQUARE KEYS

DESIGN PROBLEMS

521. A 2-in. shaft, of cold-drawn AISI 1137, has a pulley keyed to it. (a) Compute the length of square key and the length of flat key such that a key made of cold-drawn C1020 has the same yield strength as the shaft does in pure torsion. (b) The same as (a), except that the key material is AISI 2317, OQT 1000°F. (c) Would you discard either of these keys? Explain.

522. A cast-iron pulley transmits 65.5 hp at 1750 rpm. The 1045 as-rolled shaft to which it is to be keyed is $1\frac{3}{4}$ in. in diameter; key material, cold-drawn 1020. Compute the length of flat key and of square key needed.

523. A $3\frac{1}{4}$-in. shaft transmits with medium shock 85 hp at 100 rpm. Power is received through a sprocket (annealed nodular iron 60-45-10) keyed to the shaft of cold-rolled AISI 1040 (10% work), with a key of cold-finished B1113. What should be the length of (a) a square key? (b) a flat key?

524. A cast-steel gear (SAE 0030), with a pitch diameter of 36 in., is transmitting 75 hp at 210 rpm to a rock crusher, and is keyed to a 3-in. shaft (AISI 1045, as rolled); the key is made of

AISI C1020, cold drawn. For a design factor of 4 based on yield strength, what should be the length of (a) a square key, (b) a flat key? (c) Would either of these keys be satisfactory?

525. An electric motor delivers 50 hp at 1160 rpm to a $1\frac{5}{8}$-in. shaft (AISI 13B45, OQT 1100°F). Keyed to this shaft is a cast-steel (SAE 080, N&T) pulley whose hub is 2 in. long. The loading may be classified as mild shock. Decide upon a key for this pulley (material), investigating both flat and square keys.

CHECK PROBLEMS

526. A cast-steel (SAE 080, N&T) pulley, attached to a 2-in. shaft, is transmitting 40 hp at 200 rpm, and is keyed by a standard square key, 3 in. long, made of SAE 1015, cold drawn; shaft material, C1144, OQT 1000°F. (a) What is the factor of safety of the key? (b) The same as (a) except a flat key is used.

527. A cast-steel (SAE 080, N&T) pulley is keyed to a $2\frac{1}{2}$-in. shaft by means of a standard square key, $3\frac{1}{2}$-in. long, made of cold-drawn SAE 1015. The shaft is made of cold-drawn AISI 1045. If the shaft is in virtually pure torsion and turns at 420 rpm, what horsepower could the assembly safely transmit (steady loading)?

528. The same as **527,** except that the diameter is 3 in. and the length of the key is 5 in.

MISCELLANEOUS KEYS

529. Two assemblies, one with one feather key and the other with two feather keys, are shown, with the assumed positions of the normal forces N. Each assembly is transmitting a torque T. Derive an equation for each case giving the axial force needed to slide the hub along the shaft ($f =$ coefficient of friction). Does either have an advantage in this respect?

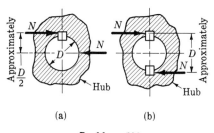

| (a) | (b) |

Problem 529.

530. A $1^{11}/_{16}$-in. shaft rotating at 200 rpm, carries a cast-iron gear keyed to it by a $\frac{1}{4} \times 1\frac{1}{4}$-in. Woodruff key; shaft material is cold-finished SAE 1045. The power is transmitted with mild shock. What horsepower may be safely transmitted by the key, (a) if it is made of cold-drawn SAE 1118? (b) if it is made of SAE 2317, OQT 1000°F? (c) How many keys of each material are needed to give a capacity of 25 hp? Specify a choice.

531. A $\frac{3}{16} \times 1$-in. Woodruff key is used in a $1\frac{3}{16}$-in. shaft (cold-drawn SAE 1045). (a) If the key is made of the same material, will it be weaker or stronger than the shaft in pure torsion? (b) If the key is made of SAE 4130, WQT 1100°F, will it be weaker or stronger? For the purposes here, the weakening of the shaft by the keyway is ignored.

532. A 2-in. shaft (cold-finished SAE 1137) is connected to a hub by a $\frac{3}{8}$-in. radial taper pin made of 4150,

OQT 1000°F. (a) What horsepower at 1800 rpm would be transmitted when the pin is about to be sheared off? (b) For this horsepower, what peak torsional stress may be repeated in the shaft? Is the shaft safe from fatigue at this stress?

533. A 20-in. lever is keyed to a $1\frac{7}{8}$-in. shaft (cold-finished SAE 1141) by a radial taper pin whose mean diameter is 0.5 in.; pin material, C1095, OQT 800°F. The load on the lever is repeatedly reversed; $N = 2$ on endurance strength. What is the safe lever load (a) for the shaft, (b) for the pin key (shear only), (c) for the combination?

534. A lever is keyed to a $2\frac{1}{2}$-in. shaft made of SAE 1035, as rolled, by a radial taper pin, made of SAE 1020, as rolled. A load of 200 lb. is applied to the lever 22 in. from the center of the shaft. (a) What size pin should be used for $N = 3$ based on the yield strength in shear? (b) Let the hub diameter be 5 in. and assume that the part of the pin in the hub is a uniformly loaded cantilever beam. Compute the bending stress and comment on the bending strength (especially if the loading varies).

535. A sprocket, transmitting 10 hp at 100 rpm, is attached to a $1\frac{7}{16}$-in. shaft as shown in Fig. 10.15, p. 290, *Text;* $E = 3\frac{1}{2}$-in. What should be the minimum shear pin diameter if the computed stress is 85% of the breaking stress mentioned in the *Text?*

536. A gear is attached to a 2-in. shaft somewhat as shown in Fig. 10.15, p. 290, *Text;* $E = 3\frac{5}{16}$ in.; minimum shear-pin diameter $= \frac{3}{8}$ in. with a rated torque of 22 in-kips. (a) For this torque, compute the stress in the shear pin. (b) From the ferrous metals given in the Appendix, select those that would give a resisting torque of about 110% of the rated value. Choose one, specifying its heat treatment or other conditions.

SPLINES

537. A shaft for an automobile transmission has 10 splines with the following dimensions: $D = 1.25$ in., $d = 1.087$ in., and $L = 1.000$ in. (see Table 10.2, p. 287, *Text*). Determine the safe torque capacity and horsepower at 3600 rpm of this sliding connection.

538. The rear axle of an automobile has one end splined. For this fitting there are ten splines, and $D = 1.31$ in., $d = 1.122$ in., and $L = 1^{15}/_{16}$ in. The minimum shaft diameter is $1^3/_{16}$ in. (a) Determine the safe torque capacity of the splined connection, sliding under load. (b) Determine the torque that would have the splines on the point of yielding if the shaft is AISI 8640, OQT 1000°F, if one fourth of the splines are in contact. (c) Determine the torsional stress in the shaft corresponding to each of these torques.

539. An involute splined connection has 10 splines with a pitch P_d of 12/24. (a) Determine the dimensions of this connection. (b) Compute the length of spline to have the same torsional strength as the shaft when one fourth the splines carry the load; minimum shaft diameter is $^9/_{16}$ in. (no sliding). Check for compression.

COUPLINGS

540. A flange coupling has the following dimensions (Fig. 10.19, p. 291, *Text*): $d = 5$, $D = 8^5/_8$, $H = 12^1/_4$, $g = 1^1/_2$, $h = 1$, $L = 7^1/_4$ in.; number of bolts = 6; $1^1/_4 \times 1^1/_4$-in. square key. Materials: key, cold-drawn AISI 1113; shaft, cold-rolled AISI 1045; bolts, SAE grade 5 (§ 5.8). Using the static approach with $N = 3.3$ on yield strengths, determine the safe horsepower that this connection may transmit at 630 rpm.

541. A cast-iron (ASTM 25) jaw clutch with 4 jaws transmits 50 hp at 60 rpm. The inside diameter of the jaws is 3 in. Considering rough handling, choose $N = 8$ on ultimate strengths. Make reasonable and conservative assumptions and compute (a) the outside diameter of the jaws, (b) the length of jaws h.

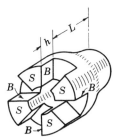

Problem 541.

542. The universal joint shown is made of AISI 3150, OQT 1000°F; $a = 2^7/_{16}$ in., $D = ^9/_{16}$; $n = 400$ rpm. (a) What torque may be transmitted for shear of the pin ($N = 5$ on ultimate)? (b) Considering the pin as a simply supported beam of length a with the load distributed from a maximum at one end to a negative maximum at the other (triangular), compute the safe transmitted torque (same N). (c) In order not to have excessive wear on the pin, the average bearing pressure should not exceed 3 ksi. Compute this transmitted torque. (d) What is the safe power?

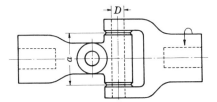

Problems 542, 543.

543. The same as **542**, except that $a = ^3/_8$ in. and $D = ^1/_8$ in.

544. A diagrammatic representation of a universal joint is shown, two yoke parts, the type being similar to Figs. 10.28 and 12.10, *Text*. The pin extensions have a diameter $D = ^3/_4$ in.; $a = ^{11}/_{16}$ in.; material of all parts is 4340, OQT 800. Let $N = 4$ on ultimate stresses; $n = 2400$ rpm. Compute the safe torque for (a) shear of pins, (b) the pin extensions in bending, assuming that the load distribution is from zero at the outside pin ends to a maximum at the inside yoke surfaces, (c) an average bearing pressure on pins of 4 ksi. (d) What is the corresponding horsepower capacity?

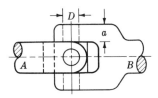

Problem 544.

545–550. These numbers may be used for other problems.

Section 9

JOURNAL AND PLANE-SURFACE BEARINGS

NOTE. *In solving 360° bearing problems with $L/D = 1$, use the charts, Figs. AF 17 and 18, as far as they will take you whenever an interpolation in the tables is necessary.*

LIGHTLY LOADED BEARINGS

551. (a) A 3×3-in. full bearing supports a load of 900 lb., $c_d/D = 0.0015$, $n = 400$ rpm. The temperature of the SAE 40 oil is maintained at 140°F. Considering the bearing lightly loaded (Petroff), compute the frictional torque, fhp, and the coefficient of friction. (b) The same as (a) except that the oil is SAE 10W.

552. The same as **551**, except that $n = 6300$ rpm.

553. The average pressure on a 6-in. full bearing is 50 psi, $c_d = 0.003$ in., $L/D = 1$. While the average oil temperature is maintained at 160°F with $n = 300$ rpm, the frictional force is found to be 13 lb. Compute the coefficient of friction and the average viscosity of the oil. To what grade of oil does this correspond?

FULL BEARINGS

554. The load on a 4-in. full bearing is 2000 lb., $n = 320$ rpm; $L/D = 1$; $c_d/D = 0.0011$; operating temperature = 150°F; $h_o = 0.00088$ in. (a) Select an oil that will closely accord with the stated conditions. For the selected oil determine

(b) the frictional loss (ft-lb./min.), (c) the hydrodynamic oil flow through the bearing, (d) the amount of end leakage, (e) the temperature rise as the oil passes through the bearing, (f) the maximum pressure.

555. A 4-in., 360° bearing, with $L/D = 1.1$ (use table and chart values for 1), is to support 5 kips with a minimum film thickness 0.0008 in.; $c_d = 0.004$ in., $n = 600$ rpm. Determine (a) the needed absolute viscosity of the oil, (b) a suitable oil if the average film temperature is 160°F, (c) the frictional loss in hp. (d) Adjusting only h_o to the optimum value for minimum friction, determine the fhp and compare. (e) This load varies. What could be the magnitude of the maximum impulsive load if the eccentricity ratio ϵ becomes 0.8? Ignore "squeeze" effect.

556. For an 8×4-in. full bearing, $c_r = 0.0075$ in., $n = 2700$ rpm, average $\mu = 4 \times 10^{-6}$ reyns. (a) What load may this bearing safely carry if the minimum film thickness is not to be less than that given by Norton, § 11.14, *Text?* (b) Compute the corresponding frictional

59

loss (fhp). (c) Complete calculations for the other quantities in Table AT 20: ϕ, q, q_s, Δt_o, p_{max}. Compute the maximum load for an optimum (load) bearing (d) if c_r remains the same, (e) if h_o remains the same.

557. A 6×6-in. full bearing has a frictional loss of fhp = 11 when the load is 68,500 lb. and $n = 1600$ rpm; $c_r/r = 0.001$. (a) Compute the minimum film thickness. Is this in the vicinity of that for an optimum bearing? (b) What is the viscosity of the oil and a proper grade for an operating temperature of 160°F? (c) For the same h_o, but for the maximum-load optimum, determine the permissible load and the fhp.

558. The maximum load on a 2.25×1.6875-in. main bearing of an automobile is 3140 lb. with wide-open throttle at 1000 rpm. If the oil is SAE 20W at 210°F, compute the minimum film thickness for a bearing clearance of (a) 0.0008 in. and (b) 0.0005 in. Which clearance results in the safer operating conditions? *Note:* Since a load of this order exists for only 20–25° of rotation, the actual h_o does not reach this computed minimum (squeeze effect).

559. The same as **558**, except that the oil is SAE 40.

560. If the bearing described in **558** wears to a clearance of 0.002 in. and if the same oil is used, are the operating conditions safer or not? Show calculations.

561. A 360° bearing supports a load of 2500 lb.; $D = 5$ in., $L = 2.5$ in., $c_r = 0.003$ in., $n = 1800$ rpm; SAE 20W oil entering at 100°F. (a) Compute the average temperature t_{av} of the oil through the bearing. (An iteration procedure. Assume μ; compute S and the corresponding Δt_o; then the average oil temperature $t_{av} = t_i + \Delta t_o/2$. If this t_{av} and the assumed μ do not locate a point in Fig. AF 16 on the line for SAE 20W oil, try again.) Calculate (b) the minimum film thickness, (c) the fhp, (d) the amount of oil to be supplied and the end leakage.

PARTIAL BEARINGS

562. A 2×2-in. bearing has a clearance $c_r = 0.001$ in. and $h_o = 0.0004$ in.; $n = 2400$ rpm, and for the oil,

$\mu = 3 \times 10^{-6}$ reyns. Determine the load, frictional horsepower, the amount of oil to enter, the end leakage of oil, and the temperature rise of the oil as it passes through for: (a) a full bearing, partial bearings of (b) 180°, (c) 120°, (d) 90°, (e) 60°.

563. A 2×2-in. bearing sustains a load of $W = 5000$ lb.; $c_r = 0.001$ in.; $n = 2400$ rpm; $\mu = 3 \times 10^{-6}$ reyns. Using Figs. AF 17 and AF 18, determine the minimum film thickness and the frictional loss (ft-lb./min.) for (a) a full bearing, and for partial bearings of (b) 180°, (c) 120°, (d) 90°, (e) 60°.

564. A 120° partial bearing is to support 4500 lb. with $h_c = 0.002$ in.; $L/D = 1$; $D = 4$ in.; $c_d = 0.010$ in.; $n = 3600$ rpm. Determine (a) the oil's viscosity, (b) the frictional loss (ft-lb./min.), (c) the eccentricity angle, (d) the needed oil flow, (e) the end leakage, (f) the temperature rise of the oil as it passes through, (g) the maximum pressure. (h) If the clearance given is the average, what approximate class of fit (Table 3.1) is it? (i) What maximum impulsive load would be on the bearing if the eccentricity ratio suddenly went to 0.8? Ignore "squeeze" effect.

565. The same as **564**, except that $L/D = 0.5$.

566. A 120° partial bearing is to support 4500 lb., $D = 3$ in. $L = 3$ in., $c_d = 0.003$ in.; $n = 3600$ rpm; SAE 20W entering at 110°F. Calculate (a) the average temperature of the oil as it passes through, (b) the minimum film thickness, (c) the fhp, (d) the quantity of oil to be supplied. HINT: In (a) assume μ and determine the corresponding values of S and Δt_o; then $t_{av} = t_i + \Delta t_o/2$. If assumed μ and t_{av} do not locate a point in Fig. AF 16 that falls on line for SAE 20W, iterate.

567. The 6000-lb. reaction on an 8×4-in., 180° partial bearing is centrally applied; $n = 1000$ rpm; $h_o = 0.002$ in. For an *optimum* bearing with minimum friction determine (a) the clearance, (b) the oil's viscosity, (c) the frictional horsepower. (d) Choose a c_d/D ratio either smaller or larger than that obtained in (a) and show that the friction loss is greater than that in the optimum bearing. Other data remain the same.

568. A 120° partial bearing supports 3500 lb. when $n = 250$ rpm; $D = 5$ in., $L = 5$ in.; $\mu = 3 \times 10^{-6}$ reyns. What are the clearance and minimum film thickness for an *optimum* bearing (a) for maximum load, (b) for minimum friction? (c) On the basis of the average clearance in Table 3.1, about what class fit is involved? Would this fit be on the expensive or inexpensive side? (d) Find the fhp for each optimum bearing.

569. The same as **568**, except that $n = 1250$ rpm.

570. A 180° partial bearing is to support 17,000 lb. with $p = 200$ psi., $n = 1500$ rpm, $h_o \approx 0.003$ in., $L/D = 1$. (a) Determine the clearance for an *optimum* bearing with minimum friction. (b) Taking this clearance as the average, choose a fit (Table 3.1) that is approximately suitable. (c) Select an oil for an average temperature of 150°F. (d) Compute fhp.

571. The reaction on a 120° partial bearing is 2000 lb. The 3-in. journal turns at 1140 rpm; $c_d = 0.003$ in.; the oil is SAE 20W at an average operating temperature of 150°F. Plot curves for the minimum film thickness and the frictional loss in the bearing against the ratio L/D, using $L/D = 0.25$, 0.5, 1, and 2. (*Note:* This problem may be worked as a class problem with each student being responsible for a particular L/D ratio.)

STEADY-STATE TEMPERATURE

572. A 180° partial bearing is subjected to a load of 12,000 lb.; $D \times L = 8 \times 8$ in., $c_r/r = 0.0015$, $h_c \approx 0.0024$ in., $n = 500$ rpm. The air speed about the bearing is expected to be in excess of 1000 fpm (on moving vehicle) and the effective radiating area is $20DL$. Determine: (a) the eccentricity factor, (b) μ reyns, (c) the frictional loss (ft-lb./min.), (d) the estimated temperature of the oil and bearing (a self-contained oil-bath unit) for steady-state operation, and a suitable oil. (e) Compute Δt_o of the oil passing through the load-carrying area, remark on its reasonableness, and decide upon whether some redesign is desirable.

573. A 2 × 2-in. full bearing (ring-oiled) has a clearance ratio $c_d/D = 0.001$. The journal speed is 500 rpm, $\mu = 3.4 \times 10^{-6}$ reyns, and $h_o = 0.0005$ in. The ambient temperature is 100°F; $A_b = 25DL$, and the transmittance is taken as $h_{cr} = 2$ Btu/hr-sq. ft.-°F. Calculate (a) the total load for these conditions, (b) the frictional loss, (c) the average temperature of the oil for steady-state operation. Is this temperature satisfactory? (d) For the temperature found, what oil do you recommend? For this oil will h_o be less or greater than the specified value? (e) Compute the temperature rise Δt_o of the oil as it passes through the bearing. Is this compatible with other temperatures found? (f) What minimum quantity of oil should the ring deliver to the bearing?

574. An 8 × 9-in. full bearing (consider $L/D = 1$ for table and chart use only) supports 15 kips with $n = 1200$ rpm; $c_r/r = 0.0012$; construction is medium heavy with a radiating-and-convecting area of about $18DL$; air flow about the bearing of 80 fpm may be counted on (nearby pulley); ambient temperature is 90°F. Decide upon a suitable minimum film thickness. (a) Compute the frictional loss and the steady-state temperature. Is additional cooling needed for a reasonable temperature? Determine (b) the temperature rise of the oil as it passes through the load-carrying area and the grade of oil to be used if it enters the bearing at 130°F, (c) the quantity of oil needed.

575. A 3.5 × 3.5-in., 360° bearing has $c_r/r = 0.0012$; $n = 300$ rpm; desired minimum $h_o \approx 0.0007$ in. It is desired that the bearing be self-contained (oil-ring); air-circulation of 80 fpm is expected; heavy construction, so that $A_b \approx 25DL$. For the first look at the bearing, assume $\mu = 2.8 \times 10^{-6}$ reyns and compute (a) the frictional loss (ft-lb./min.), (b) the average temperature of the bearing and oil as obtained for steady-state operation, (c) Δt_o as the oil passes through the load-carrying area (noting whether comparative values are reasonable). (d) Select an oil for the steady-state temperature and decide whether there will be any overheating troubles.

576. A 10-in. full journal for a steam-turbine rotor that turns 3600 rpm supports a 20-kip load with $p = 200$ psi; $c_r/r = 0.00133$. The oil is to have $\mu = 2.06 \times 10^{-6}$ reyns at an average oil temperature of 130°F. Compute (a) the minimum film thickness (comment on its adequacy), (b) the fhp, (c) the attitude angle, the maximum pressure, and the quantity of oil that passes through the load-carrying area (gpm). (d) At what temperature must the oil be introduced in order to have 130°F average? (e) Estimate the amount of heat lost by natural means from the bearing (considered oil bath) with air speed of 300 fpm. If the amount of oil flow computed above is cooled back to the entering temperature, how much heat is removed? Is this total amount of heat enough to care for frictional loss? If not, what can be done (§ 11.21)?

577. A 3 × 3-in. full bearing is to support 785 lb. at 360 rpm. The oil, supplied at low pressure at 100°F, is a light turbine oil, and is circulated to carry away the heat. Plot curves showing the temperature rise of the oil, the minimum film thickness, and the coefficient of friction against diametral clearance for values of c_d from 0.001 in. to 0.007 in. (HINT: A trial-and-error solution is required.) For each clearance, assume an oil viscosity, calculate S, determine Δt_o from Table AT 20; μ is based on the average oil-film temperature t_{av}, approximately $t_i + \Delta t_o/2$. Repeat until the point for μ and t_{av} falls on line for light turbine oil, Fig. AF 16. (*Note:* This may be worked as a class problem, with each student assigned a particular value of c_d.)

DESIGN PROBLEMS

578. A 3.5-in. full bearing on an air compressor is to be designed for a load of 1500 lb.; $n = 300$ rpm; let $L/D = 1$. Probably a medium running fit would be satisfactory. Design for an average clearance that is decided by considering both Tables 3.1 and 11.1. Choose a reasonable h_o, say one that gives $h_o/c_r \approx 0.5$. Compute all parameters that are available via the *Text* after you have decided on the details. It is desired that the bearing

operate at a reasonable steady-state temperature (perhaps ring-oiled, medium construction), without special cooling. Specify the oil to be used and show all calculations to support your conclusions. What could be the magnitude of the maximum impulsive load if the eccentricity ratio ϵ becomes 0.8, "squeeze" effect ignored?

579. Because there is some concern that space may become a problem in the air compressor of **578**, it has been decided to investigate the possibility of using $L/D = 0.5$, otherwise this problem is the same as **578**.

580. A 2500-kva generator, driven by a water wheel, operates at 900 rpm. The weight of the rotor and shaft is 15,100 lb. The left-hand, 5-in., full bearing supports the larger load, $R = 8920$ lb. The bearing should be about medium-heavy construction (for estimating A_b). (a) Decide upon an average clearance considering both Tables 3.1 and 11.1, and upon a minimum film thickness ($h_o/c_r \approx 0.5$ is on the safer side). (b) Investigate first the possibility of the bearing being a self-contained unit without need of special cooling. Not much air movement about the bearing is expected. Then make final decisions concerning the oil, clearance, and film thickness and compute all the parameters given in the *Text*, being sure that everything is reasonable.

PRESSURE FEED

581. An 8 × 8-in. full bearing supports 5 kips at 600 rpm of the journal; $c_r = 0.006$ in.; let the average $\mu = 2.5 \times 10^{-6}$ reyns. (a) Compute the frictional loss U_f. (b) The oil is supplied under a 40-psi gage pressure with a longitudinal groove at the point of entry. Assuming that other factors, including U_f, remain the same and that the heat loss to the surroundings is negligible, determine the average temperature rise of the circulating oil.

582. The same as **581**, except that there is a circumferential groove dividing the bearing into two 4-in. lengths.

583. A 4-in., 360° bearing, with $L/D = 1$, supports 2.5 kips with a minimum film of $h_o = 0.0008$ in., $c_d = 0.01$

in., n = 600 rpm. The average temperature rise of the oil is to be about 25°. Compute the pressure at which oil should be pumped into the bearing if (a) all bearing surfaces are smooth, (b) there is a longitudinal groove at the oil-hole inlet. (c) The same as (a), except that there is a 360° circumferential groove dividing the bearing into 2-in. lengths.

BEARING CAPS

584. An 8-in. journal, supported on a 150° partial bearing, is turning at 500 rpm; bearing length = 10.5 in., c_d = 0.0035 in., h_o = 0.00106 in. The average temperature of the SAE 20W oil is 170°F. Estimate the frictional loss in a 160° cap for this bearing.

585. A partial 160° bearing has a 160° cap; D = 2 in., L = 2 in., c_d = 0.002 in., h_o = 0.0007 in. n = 500 rpm, and μ = 2.5 × 10^{-6} reyns. For the cap only, what is the frictional loss?

586. The central reaction on a 120° partial bearing is 10 kips; D = 8 in., L/D = 1, c_r/r = 0.001. Let n = 400 rpm and μ = 3.4 × 10^{-6} reyns. The bearing has a 150° cap. (a) For the bearing and the cap, compute the total frictional loss by adding the loss in the cap to that in the bearing. (b) If the bearing were 360°, instead of partial, calculate the frictional loss and compare.

587. The central reaction on a 120° partial bearing is 10 kips; D = 8 in., L/D = 1, c_r/r = 0.001; n = 1200 rpm. Let μ = 2.5 × 10^{-6} reyns. The bearing has a 160° cap. (a) Compute h_o and fhp for the bearing and for the cap to get the total fhp. (b) Calculate the fhp for a full bearing of the same dimensions and compare. Determine (c) the needed rate of flow into the bearing, (d) the side leakage q_s, (e) the temperature rise of the oil in the bearing both by equation (o), § 11.13, *Text*, and by Table AT 22. (f) What is the heat loss from the bearing if the oil temperature is 180°F? Is the

natural heat loss enough to cool the bearing? (g) It is desired to pump oil through the bearing with a temperature rise of 12°F. How much oil is required? (h) For the oil temperature in (f), what is a suitable oil to use?

IMPERFECT LUBRICATION

588. A 0.5 × 0.75-in. journal turns at 1140 rpm. What maximum load may be supported and what is the frictional loss if the bearing is (a) SAE Type I, bronze base, sintered bearing, (b) nylon (Zytel) water lubricated, (c) Teflon, with intermittent use, (d) one with carbon-graphite inserts.

589. The same as **588**, except that n = 375 rpm.

590. A bearing to support a load of 150 lb. at 800 rpm is needed; D = 1 in.; semilubricated. Decide upon a material and length of bearing, considering sintered metals, Zytel, Teflon, and graphite inserts.

591. The same as **590**, except that the load is 500 lb., and n = 100 rpm.

THRUST BEARINGS

592. A 4-in. shaft has on it an axial load of 8000 lb., taken by a collar thrust bearing made up of five collars, each with an outside diameter of 6 in. The shaft turns 150 rpm. Compute (a) the average bearing pressure, (b) the approximate work of friction.

593. A 4-in. shaft, turning at 175 rpm, is supported on a step bearing. The bearing area is annular, with a 4-in. outside diameter and a ¾-in. inside diameter. Take the allowable average bearing pressure as 180 psi. (a) What axial load may be supported? (b) What is the approximate work of friction?

594–600. These numbers may be used for other problems.

Section 10

BALL AND ROLLER BEARINGS

NOTE. *If bearing manufacturers' catalogs are available, the student should make at least two solutions to each assigned problem using different catalogs.*

CONSTANT LOADING

601. The radial reaction on a bearing is 1500 lb.; it also carries a thrust of 1000 lb.; shaft rotates 1500 rpm; outer ring stationary; smooth load, 8-hr./day service, say 15,000 hr. (a) Select a deep-groove ball bearing. (b) What is the rated 90% life of the selected bearing? (c) For $b = 1.34$, compute the probability of the selected bearing surviving 15,000 hr.

602. A certain bearing is to carry a radial load of 500 lb. and a thrust of 300 lb. The load imposes light shock; the desired 90% life is 10 hr./day for 5 years at $n = 3000$ rpm. (a) Select a deep-groove ball bearing. What is its bore? Consider all bearings that may serve. (b) What is the computed rated 90% life of the selected bearing? (c) What is the computed probability of the bearing surviving the specified life? (d) If the loads were changed to 400 and 240 lb., respectively, determine the probability of the bearing surviving the specified life, and the 90% life under the new load.

603. The smooth loading on a bearing is 500-lb. radial, 100-lb. thrust; $n = 300$ rpm. An electric motor drives through gears; 8 hr./day, fully utilized.

(a) Considering deep-groove ball bearings that may serve, choose one and specify its bore. For the bearing chosen, determine (b) the rated 90% life and (c) the probability of survival for the design life.

604. The same as **603**, except that the service is to be continuous, 24 hr./day.

605. A No. 311, single-row, deep-groove ball bearing is used to carry a radial load of 1500 lb. at a speed of 500 rpm. (a) What is the 90% life of the bearing in hours? What is the approximate median life? What is the probability of survival if the actual life is desired to be (b) 10^5 hr., (c) 10^4 hr.?

606. The load on an electric-motor bearing is 350 lb., radial; 24-hr. service, $n = 1200$ rpm; compressor drive; outer race stationary. (a) Decide upon a deep-groove ball bearing, giving its significant dimensions. Then compute the selected bearing's 90% life, and the probable percentage of failures that would occur during the design life. What is the approximate median life of this bearing? (b) The same as (a), except that a 200 series roller bearing is to be selected.

64

607. The same as **606,** except that the radial load is 650 lb.

608. A deep-groove ball bearing on a missile, supporting a radial load of 200 lb., is to have a design life of 20 hr.; with only a 0.5% probability of failure while $n = 4000$ rpm. Using a service factor of 1.2, choose a bearing. (A 5- or 6-place log table is desirable.)

609. The same as **608,** except that it has been decided to have a probability of survival of 0.999.

VARIABLE LOADS

610. A certain bearing is to carry a radial load of 10 kips at a speed of 10 rpm for 20% of the time, a load of 8 kips at a speed of 50 rpm for 50% of the time, and a load of 5 kips at 100 rpm during 30% of the time, with a desired life of 3000 hr.; no thrust. (a) What is the cubic mean load? (b) What ball bearings may be used? (c) What roller bearings?

611. The same as **610,** except that the speed remains constant at 50 rpm.

612. A deep-groove ball bearing No. 215 is to operate 30% of the time at 500 rpm with $F_x = 1200$ lb. and $F_z = 600$ lb., 55% of the time at 800 rpm with $F_x = 1000$ lb. and $F_z = 500$ lb., and 15% of the time at 1200 rpm with $F_x = 800$ lb. and $F_z = 400$ lb. Determine (a) the cubic mean load; (b) the 90% life of this bearing in hours, (c) the average life in hours.

613. The same as **612,** except that the speed remains constant at 800 rpm.

MANUFACTURER'S CATALOGS NEEDED

614. A shaft for the general-purpose gear-reduction unit described in **489** has radial bearing reactions of $R_C = 613$ lb. and $R_D = 1629$ lb.; $n = 250$ rpm. As-sume that the unit will be fully utilized for at least 8 hr./day, with the likelihood of some uses involving minor shock. (a) Select ball bearings for this shaft. (b) Select roller bearings. (c) What is the probability of both bearings C and D surviving for the design life?

615. A shaft similar to that in **478** has the following radial loads on the bearings, left to right: 803 lb., 988 lb., 84 lb., and 307 lb.; no thrust. The minimum shaft diameters at the bearings are: 1.250 in., 1.125 in., 1.000 in., and 1.0625 in. Assume that the service will not be particularly gentle; intermittently used, with $n = 425$ rpm. (a) Select ball bearings for this shaft. (b) Select roller bearings. (c) Compare your solution with the one for the same installation as given in the *Timken Engineering Journal.*

616. The shaft described in **492** is supported by double-row ball bearings: No. 5312 at A with $F_x = 1830$ lb. (no thrust); No. 5315 at G with $F_x = 3680$ lb. and $F_z = 598$ lb.; assume fully utilized 8-hr./day service at $n = 388$ rpm. (a) Determine the expected 90% life, (b) the probability of a particular bearing at each location surviving the design life, (c) the probability of both bearings surviving the design life. (d) If the decision were yours, what bearings would you recommend?

617. In accordance with the instructor's choice, select ball or roller bearings as alternate choices for a problem that you have already solved as a journal bearing. Use catalogs if available.

618. In accordance with instructor's choice, select ball or roller bearings to support the shaft for a shaft problem that you have already solved. Use catalogs if available.

619–630. These numbers may be used for other problems.

Section 11

SPUR GEARS

DESIGN PROBLEMS

631. A pair of gears with 20° full-depth teeth are to transmit 10 hp at 1750 rpm of the 3-in. pinion; velocity ratio desired is about 3.8; intermittent service. Use a strength reduction factor of about 1.4, with the load at the tip and teeth commercially cut. Determine the pitch, face width, and tooth numbers if the material is cast iron, class 20.

632. Since there is some interference in the design of **631**, it is decided to try accurately cut teeth or an annealed cast-steel (SAE 0030) pinion or both. Investigate these alternatives and make a recommendation, giving reasons for your decisions.

633. A pair of gears with 20° full-depth teeth are to transmit 5 hp at 1800 rpm of the pinion; $m_\omega = 2.5$; $N_p = 18$ teeth; commercially cut teeth; intermittent service; $K_f \approx 1.45$. (a) Determine the pitch, face width, and tooth numbers if the material is cast iron, class 25. (b) The same as (a), except that the pinion is to be made of phosphor gear bronze (SAE 65, Table AT 3).

634. It is desired to transmit 120 hp at 1800 rpm of the pinion; intermittent service; with light shock (§ 13.18); prefer-

ably not less than 18, 20°-full-depth teeth on the pinion; $K_f = 1.5$ should be conservative; $m_\omega = 1.5$. Decisions must be made concerning the material and quality of cutting the teeth. Since the design is for strength only, it will be convenient to express F_t, F_d, v_m, b in terms of P_d and arrange an equation containing s and P_d convenient for iteration. Weak material results in a relatively large pinion with high peripheral speed. A very strong material may be unnecessarily expensive. On a production basis, carefully cut teeth should have a reasonable cost. Specify material, accuracy of cutting, pitch, face width, and tooth numbers.

635. The same as **634,** except that 80 hp is transmitted at 1200 rpm of the pinion.

CHECK PROBLEMS

636. A pair of carefully cut, full depth, 20° involute gears, made of cast iron, ASTM 30, is transmitting 5 hp at 1150 rpm of the pinion; $N_p = 24$, $N_g = 32$, $P_d = 8$, $b = 1\frac{1}{2}$ in. For the teeth, determine (a) the endurance strength (b) the dynamic load, (c) the service factor (§ 13.18).

637. A manufacturer's catalog for cut-tooth spur gears rates a 25-tooth,

66

cast-iron (ASTM 25) pinion with 5 pitch, 20° full-depth involute teeth at 16.5 hp at 900 rpm; $b = 2\frac{1}{2}$ in. and $m_g = 2$; let $K_f = 1.5$; intermittent service; smooth load. (a) What horsepower may these gears transmit? Do you consider the catalog rating too high or too low? (b) The same as (a) except that the teeth are carefully cut. (c) The same as (a) except that the pinion is to be made of phosphor bronze (SAE 65, Table AT 3).

638. A pair of commercially cut spur gears transmits 10 hp at 1750 rpm of the 25-tooth pinion. The teeth are 20° full depth with 6 pitch; material, cast iron, class 30; face width is $1\frac{9}{16}$ in.; $N_g = 40$. Allow for stress concentration. (a) Compute the service factor for the teeth (§ 13.18). (b) If the drive is for a single-cylinder compressor, would carefully cut teeth be advisable? Show calculations.

CONTINUOUS SERVICE

DESIGN PROBLEMS

NOTE. *When using Buckingham's F_d equation and a K_f is used, as intended in design for continuous service, use Y for load near middle.*

639. The pinion of a pair of steel gears, transmitting 110 hp at 2300 rpm, is to have a diameter of about $4\frac{1}{3}$ in.; $m_g \approx 2.3$; 20° full-depth teeth; the drive is to a centrifugal pump, continuous service. (a) Decide upon P_d, b, N_p, N_g, and the material to be used. Consider the strength with the load near the middle of the profile. (b) The same as (a) except that it is not expected that the maximum loading will occur for more than 10^7 cycles, can you justify changes in your previous answers?

640. A pinion with 20° full-depth teeth, transmitting 160 hp at 2400 rpm, is part of a gear reduction for a lobe blower. It is to be about 3.2 in. in diameter; $m_g \approx 1.56$. (a) Decide upon a material for the mating gears (and its treatment), P_d, b, N_p, and N_g. Determine the strength with the load near the middle of the profile. (b) The same as (a), except that the maximum loading will occur for no more than 10^7 cycles.

641. Gears with 20° full-depth teeth are to transmit 100 hp continuously at 5000 rpm with $m_g = 4$; pinion $D_p = 3$ in.; the drive is subjected to minor shocks with frequent starts. First calculations are to be made for carburized pinion teeth of AISI E3310, SOQT 450°F, and the gear of cast steel, SAE 0175, WQT. Decide upon P_d, b, N_p, and N_g.

642. A 20-tooth (20° F.D.) pinion is to transmit 50 hp at 600 rpm, the service being indefinitely continuous in a conveyor drive; $m_\omega = 2.5$. The original plan is to use a nodular-iron casting, 80-60-03, for each gear. Determine suitable values for the pitch, face width, and diameters. (Warning: compute C.)

643. A 4.8-in. (approximate) pinion with 20° full-depth teeth is to transmit 40 hp at 1000 rpm; indefinitely continuous service with smooth load; $m_g = 3.5$; carefully cut teeth to reduce the dynamic load. To reduce the chance of an explosive spark, the use of a phosphor-gear-bronze (SAE 65, Table AT 3) pinion and a cast-iron (class 35) gear is a tentative decision. Decide upon an appropriate P_d and b, using Buckingham's average dynamic load.

644. The 20° full-depth teeth for a pair of steel gears are to transmit 40 hp at 1200 rpm of the 20-tooth pinion; $m_g = 3$; continuous service and indefinite life. The driven machine is an off-and-on reciprocating compressor. (a) Determine the pitch, face width, and steel (with treatment), considering at least three alternatives, including carefully cut teeth. For the gear teeth decided on, what would be the power capacity if only intermittent service (wear not considered) were required? (c) If a limited life of 10^7 cycles were satisfactory?

645. A pair of spur gears, delivering 100 hp to a reciprocating pump at a pinion speed of 600 rpm, is to serve continuously with indefinite life; minimum number of 20° full-depth teeth is 18; $m_\omega = 2.5$. Since low weight is highly important, it is decided that the initial design be for carburized case-hardened teeth. (a) Determine a suitable pitch, face width, diameters, and specify material and its heat treatment. (b) Use the same size teeth as determined in (a), but let the material be flame-hardened 4150, OQT 1100°F. Compute F_s and F_w. If it were decided that the maximum (specified) loading would be imposed only occasionally, would these gears transmit more or less power than the carburized teeth? Explain.

CHECK PROBLEMS

646. A 6-ft. ball mill runs at 24.4 rpm, the drive being through $14\frac{1}{2}°$ involute spur gears; $P_d = 2$, $N_p = 15$, $N_g = 176$, $b = 5$ in., and hp = 75. The material of the pinion is SAE 1040, BHN = 180; of the gear, 0.35% C cast steel, BHN = 180. (a) Check for strength and wear and give your decision as to the service to be expected. (b) The foregoing pinion wore out. Actually, the first step was to replace it with one made of SAE 3140, OQT 1000°F. Would you expect this to cure the trouble? (c) The drive in (b) also wore out. The following solution which maintained the same gear diameters, pitch and face, was proposed: 20° full-depth teeth; pinion of SAE 3140 with BHN = 350; gear of SAE 1045 with BHN = 280. Would you predict that these gears will give long service? What are the approximate tempering temperatures to get the specified hardnesses?

647. A 22-tooth pinion, transmitting 110 hp at 2300 rpm, drives a 45-tooth gear, both steel; 20° full depth; $P_d = 5$, $b = 1.5$ in.; The manufacturing process is expected to result in a maximum effective error of $e = 0.0016$ in. (a) Compute Buckingham's average dynamic load. Compute F_s and F_w if the material is (b) case-carburized AISI 8620, DOQT 300°F, (c) AISI 8742, OQT 950°F, (d) induction-hardened AISI 8742. (e) Suppose your company carries a stock of the foregoing materials. For a minimum-service factor of 1.2, which material do you recommend for (i) intermittent service, (ii) indefinitely continuous service, (iii) cycles of loading not to exceed 10^7?

648. Two mating steel gears have 16 and 25 teeth, respectively, 20° F.D.; $b = 2$ in., $P_d = 5$; pinion speed, 2400 rpm. The maximum effective error in the profiles is planned to be 0.0012 in. The drive is for a heavy-duty conveyer, continuous service. Compute and specify a reasonable rated horsepower if the gear teeth are: (a) Case carburized AISI 8620, SOQT 300°F, (b) AISI 8742, OQT 950°F and then flame-hardened, (c) AISI 8742, OQT 800°F.

649. The data for a pair of gears are: 20° F.D. teeth, $b = 1\frac{3}{4}$-in., $P_d = 6$, $N_p = 26$, $N_g = 60$, $n_p = 2300$ rpm; as-rolled AISI 1050; carefully cut teeth; $N_{sf} = 1.2$. (a) Strength alone considered, find the horsepower that may be transmitted. (b) Determine the required surface hardness in order for $F_w = F_d$, and specify a treatment that would make the gears long lasting in continuous service.

650. Gears with carefully cut, 20° F.D. teeth have $P_d = 5$, $b = 2$, $m_g = 5$, $N_p = 24$. Pinion material is manganese gear bronze (heading of Table AT 3); gear is cast iron, class 25. Gear speed $n_g = 200$ rpm; smooth load. They are to transmit 18 hp. (a) Are the teeth strong enough for intermittent service? (b) Does the limiting wear load indicate long life? *Suggestion:* Compute C for equation (13.7).

651. A 20-tooth pinion, 20° F.D., drives a 100-tooth gear. The pinion is made of SAE 1035, heat treated to Rockwell C15; the gear is cast iron class 35, HT; $P_d = 3$, $b = 2.5$ in.; carefully cut teeth; pinion speed $n_p = 870$ rpm, smooth load. (a) For a continuous service, indefinite life, what is a safe horsepower? (b) For intermittent service (wear unimportant), compute the safe horsepower.

652. A pair of steel gears is defined by $P_d = 8$, $b = 1.5$ in., $N_p = 25$, $N_g = 75$, $e = 0.001$ in., 20° F.D. If these gears may transmit continuously and without failure 75 hp at 1140 rpm of the pinion, what horsepower would be satisfactory for $n_p = 1750$ rpm?

653. The same as **652**, except that hp for $n_p = 870$ rpm is desired.

654. A gear manufacturer recommends that the following gears can transmit 25 hp at 600 rpm of the pinion during continuous 24-hr. service, indefinite life, moderate shock: $N_p = 31$, $N_g = 70$, $b = 3.25$ in., $P_d = 6$, 20° F.D.; pinion material is SAE 2335 with BHN = 300; gear material is SAE 1040 with BHN = 250. At what horsepower would you rate them?

NONMETALLIC GEARS

655. A 4-in. Bakelite pinion meshing with a cast-iron gear, is to be on the shaft of a 12-hp induction motor that turns at 850 rpm; $m_\omega = 5$, 20° F.D. teeth. (a) Determine P_d, b, N_p, N_g for indefinitely continuous service with a smooth load. Is there serious interference in the gears you have designed? (b) Are the teeth of your design strong enough to take without damage an occasional 60% overload?

656. A Zytel pinion with molded teeth is to transmit 0.75 hp to a hardened-steel gear; $n_p = 1750$ rpm, $D_p \approx 1.25$ in. Determine the pitch, face, and number of teeth on the pinion for intermittent service.

657. A 10-in. Textolite pinion, driving a hardened steel gear, transmits power at 400 rpm; $P_d = 2.5$, $b = 5$ in., $14\frac{1}{2}°$ F.D. teeth. Determine the safe horsepower (a) for smooth, continuous, indefinite service, and also (b) for limited-life intermittent service.

658. A 16-pitch Zytel pinion, with 26, 20° F.D., cut teeth, rotates at 600 rpm; $b = \frac{5}{8}$ in. (a) What safe horsepower may be transmitted for long life? (b) For 10^7 cycles?

CAST-TOOTH GEARS

659. A pair of cast-iron spur gears, ASTM 20, with cast teeth, transmits 10 hp at 125 rpm of the pinion; $m_\omega = 5$, $D_p \approx 8$ in. Determine P_c, b, N_p, N_g.

660. Design the cast teeth for a pair of cast-iron spur gears to transmit 35 hp at 50 rpm of the pinion; $m_\omega \approx 2.5$. Decide upon a suitable grade of cast iron and find P_c, b, D_p, D_g, and center distance.

661. A manufacturer's catalog specifies that a pair of gray cast-iron spur gears with cast teeth will transmit 7.01 hp at a pitch-line speed of 100 fpm; $N_p = 20$, $P_c = 1.5$ in., $b = 4$ in. Compute the stress and specify the grade of cast iron that should be used.

ARMS AND RIMS

662. A 24-in. cast-iron gear transmits 30 hp at 240 rpm; $14\frac{1}{2}°$ F.D. teeth, $P_d = 4$, $b = 2.5$ in. The gear is on a $2\frac{1}{4}$-in. shaft. Determine (a) the hub diameter, rim thickness, and bead, (b) the dimensions of the arms at the hub and at the pitch circle for an elliptical-shaped

section, (c) the arm dimensions for a cross shape.

663. A 21-in. cast-iron gear transmits 25 hp at 275 rpm. The pitch of the 20° F.D. teeth is 5, $b = 2\frac{1}{2}$ in. The shaft is $1\frac{3}{4}$ in. in diameter. Parts (a), (b), and (c) are same as in **662.**

INTERNAL GEARS

664. A 20-tooth pinion, with 20° F.D. teeth, drives a 75-tooth internal gear ($P_d = 8$, $b = 1.5$ in., $n_p = 1150$ rpm; material, cast iron, class 20). What horsepower may be transmitted continuously (a) if the teeth are commercially hobbed (AGMA equation, § 13.15), (b) if they are precision cut? There are minor irregularities in the loading.

665. The same as **664**, except that $P_d = 3.5$ and $b = 3$ in.

666. A planetary gear train is composed of four gears—the sun gear B, two planet gears C, and a fixed internal gear D, as shown. Gears B and C have 20 teeth each, gear D has 60; $P_d = 10$, $b = 1\frac{1}{4}$-in., 20° F.D. teeth, cast iron, class 20, $n_B = 1750$ rpm. (a) Determine the speed of the arm. (b) What horsepower may be transmitted continuously? Note that the dynamic load (AGMA equation, § 13.15) depends on the speed of tooth engagement, which is not the absolute pitch line speed of B (pitch-line speed relative to arm for B-C). Check the speed of tooth engagement of both B-C and C-D. (c) If the designer wishes to increase the power transmitted by using

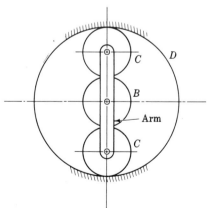

Problems 666, 667.

three planet gears, instead of two, what changes must be made in tooth numbers so that the gears can be assembled with the planets 120° apart?

667. The same as **666,** except that there are 4 planets and, for all, $P_d = 8$.

DESIGN PROJECTS

668–691. (Instructor: assign one or more students to each problem as desired.) A gear-reduction unit is to be designed according to the data in the table and the following specifications. The velocity ratio may be varied by an amount necessary to have whole tooth numbers. The given center distance is the permissible *maximum* (but this does not preclude asking the engineer in charge if a slightly larger one can be tolerated, in case it looks impossible to satisfy this condition). The teeth are to be 20° F.D. with $N_p \geq 18$ teeth, with 17 as the minimum acceptable. The service is continuous, with indefinite life. Use Buckingham's dynamic-load for average gears.

(a) Decide upon the material with its treatment, pitch, and face width. Start out being orderly with your calculations so that you do not need to copy all of them for your report. The report should show calculations for the final decisions first, but all significant calculations should be in the appendix. These latter calculations should show: that a cheap material (as cast iron) cannot be used; that through-hardened steel (minimum permissible tempering temperature is 800°F), flame- or induction-hardened steel, and carburized case-hardened teeth have all been considered in detail.

(b) To complete the design of the gears, a shaft size is needed. At the option of the instructor: (i) compute shaft diameters for pure torsion only using a conservative design stress, as $s_s = 6$ ksi (to cover stress concentration, minor bending, on the assumption that the bearings will be quite close to the gears, etc.); or (ii) make a tentative assumption of the distance between bearings, and design the shafts by a rational procedure. It would be logical for the input and output to be via flexible couplings. Let the shaft material be cold-finished AISI 1137. Design the keys for cold-drawn AISI C1118. Use better materials than these only for good reason.

Prob. No.	hp	m_g	Max. Ctr. Dist., in.	Rpm, Pinion	Kind of Load Min. N_{sf}
668	45	4	$6\frac{1}{2}$	1250	
669	40	$3\frac{1}{3}$	$6\frac{1}{2}$	800	
670	110	4	$9\frac{1}{2}$	720	Smooth
671	80	$3\frac{1}{2}$	$5\frac{1}{2}$	1800	1.1
672	30	4	$6\frac{3}{4}$	600	
673	60	4	7	900	
674	160	3	9	1100	
675	70	$3\frac{1}{2}$	$6\frac{3}{4}$	900	Minor
676	45	2	$5\frac{1}{4}$	720	Pulsations
677	80	4	9	720	
678	40	2	$4\frac{1}{2}$	820	1.2
679	90	4	$7\frac{1}{2}$	1800	
680	85	4	9	720	
681	45	3	6	1000	Mild
682	70	$2\frac{1}{2}$	$5\frac{1}{4}$	1200	Shock
683	120	3	8	1100	
684	100	6	12	1800	1.3
685	40	4	7	800	
686	70	2	$6\frac{1}{2}$	360	
687	90	5	$9\frac{1}{2}$	900	
688	55	5	$7\frac{1}{2}$	1100	Shock
689	30	4	$6\frac{1}{2}$	650	
690	110	5	12	900	1.4
691	45	$4\frac{1}{2}$	7	1260	

(c) Determine the dimensions of the hub, arms or webs, and rims and beads of both gears.

(d) Make a sketch of each gear (on separate sheets of paper) including on it all dimensions and information for its manufacture.

(e) At the instructor's option (i) choose rolling type bearings, or (ii) design sleeve bearings.

(f) Decide upon all details of the housing to enclose the gears, with sketches depicting them.

(g) Your final report should be arranged as follows: (1) title page; (2) a summary of final design decisions, computed dimensions, and material specifications; (3) sketches; (4) final calculations; (5) other calculations.

692–700. These numbers may be used for other problems.

HELICAL GEARS

DESIGN PROBLEMS

701. For continuous duty in a speed reducer, two helical gears are to be rated at 7.4 hp at a pinion speed of 1750 rpm; $m_\omega \approx 2.75$; the helix angle 15°; 20° F.D. teeth in the normal plane; let $N_p = 21$ teeth, and keep $b < 2D_p$. Determine the pitch, face, N_g, and the material and heat treatment. Use through-hardened teeth with a maximum of 250 BHN (teeth may be cut after heat treatment).

702. The same as **701**, except that the teeth are 20° F.D. in the diametral plane.

703. A pair of helical gears, subjected to heavy shock loading, is to transmit 50 hp at 1750 rpm of the pinion; $m_\omega = 4.25$; $\psi = 15°$; minimum $D_p = 4\frac{3}{4}$ in.; continuous service, 24 hr./day; 20° F.D. teeth in the normal plane, carefully cut; through-hardened to a maximum BHN = 350. Decide upon the pitch, face width, material and its treatment.

704. The same as **703**, except that the teeth are to be carburized and case hardened.

705. Design the teeth for two herringbone gears for a single-reduction speed reducer with $m_\omega = 3.80$. The capacity is 36 hp at 3000 rpm of the pinion; $\psi = 30°$; F.D. teeth with $\phi_n = 20°$. Since space is at a premium, the initial design is for $N_p = 15$ teeth and carburized teeth of AISI 8620; preferably $b < 2D_p$.

706. The same as **705**, except that $m_\omega = 4.71$ and hp = 23.

CHECK PROBLEMS

707. The data for a pair of carefully cut helical gears are: $P_{dn} = 5$, $\phi_n = 20°$, $\psi = 12°$, $b = 3.5$ in., $N_p = 18$, $N_g = 108$ teeth; pinion turns 1750 rpm. Materials: pinion, SAE 4150, OQT to BHN = 350; gear, SAE 3150, OQT to BHN = 300. Operation is with moderate shock for 8 to 10 hr./day. What horse-power may be transmitted continuously?

708. Two helical gears are used in a single-reduction speed reducer rated at 27.4 hp at a motor speed of 1750 rpm; continuous duty. The rating allows an occasional 100% momentary overload. The pinion has 33 teeth, $P_{dn} = 10$, $b = 2$ in., $\phi_n = 20°$, $\psi = 15°$; $m_\omega \approx 2.82$. For both gears, the teeth are carefully cut from SAE 1045 with BHN = 180. Compute (a) the dynamic load, (b) the endurance strength; estimate $K_f = 1.7$. Also decide whether or not the 100% overload is damaging. (c) Are these teeth suitable for continuous service? If they are not

71

suitable suggest a cure. (The gears are already cut.)

709. Two helical gears are used in a speed reducer whose input is 100 hp at 1200 rpm from an internal combustion engine. Both gears are made of SAE 4140, with the pinion heat treated to a BHN 363–415, and the gear to 321–363; let the teeth be F.D.; 20° pressure angle in the normal plane; carefully cut; helix angle $\psi = 15°$; $N_p = 22$, $N_g = 68$ teeth; $P_d = 5$, $b = 4$ in. Calculate the dynamic load, the endurance-strength load, and the limiting wear load for the teeth. Should these gears have long life if they operate continuously? (*Data courtesy of the Twin Disc Clutch Co.*)

CROSSED HELICALS

710. Helical gears are to connect two shafts that are at right angles ($N_1 = 20$, $N_2 = 40$, $P_{dn} = 10$, $\psi_1 = \psi_2 = 45°$). Determine the center distance.

711. The same as **710**, except that $\psi_1 = 60°$ and $\psi_2 = 30°$.

712. Two shafts that are at right angles are to be connected by helical gears. A tentative design is to use $N_1 = 20$, $N_2 = 60$, $P_{dn} = 10$, and a center distance of 6 in. What must be the helix angles?

DESIGN PROJECTS

713–737. Design a single-reduction helical-gear speed reducer using the data in the accompanying table. Select appropriate materials and heat treatments. Although the illustration shows a double-reduction unit, it may be helpful in deciding upon some details.

(a) For the teeth: helix angle $\psi = 15°$, F.D. with $\phi_n = 20°$; $b < 2D_p$; let the pinion have some 20 to 30 teeth. Decide upon the pitch and face width.

(b) Determine shaft sizes considering both bending and torsion. Power is supplied to the high-speed shaft through a flexible coupling. (This shaft must extend beyond the bearing far enough to mount the coupling—check catalogs for dimensions.) Power is delivered to a belt or chain that may exert the specified overhung load at a distance X (see data) from

Prob. No.	hp	m_ω	Pinion, Rpm	Over-hung Load	X, in.
713	49	7	1750	2200	
714	41	6	1150	2180	
715	44	5	870	2090	3
716	50	4	720	1930	
717	59	3	580	2160	
718	92	7	1750	2670	
719	79	6	1150	2570	
720	83	5	870	2390	3½
721	95	4	720	2115	
722	112	3	580	1850	
723	151	7	1750	3600	
724	134	6	1150	3380	
725	135	5	870	3210	4
726	158	4	720	2930	
727	187	3	580	2610	
728	227	7	1750	4480	
729	190	6	1150	4230	
730	200	5	870	4000	4½
731	228	4	720	3710	
732	270	3	580	3280	
733	300	7	1750	6460	
734	270	6	1150	6220	
735	270	5	870	5890	5
736	310	4	720	6170	
737	370	3	580	5440	

Problems 713-737.

the outside of the housing. (Note that X is not the distance to the center of the bearing.) The load should be taken in the direction that will produce the maximum bending in the shaft, because the shaft must be safe for any direction of load or rotation.

(c) Select ball or roller bearings for a 90% life of 10,000 hr. The direction of the overhung load may not be the same for maximum bearing reactions as for the maximum bending. Consider both directions of rotation. Design housing for these bearings, providing seals where the shafts extend outside the housing.

(d) Determine the dimensions of the hubs, arms or webs, rims, and beads of both gears. It may be desirable to make the pinion solid and integral with the shaft. Determine the dimensions of flat keys as needed.

(e) Make detailed sketches of each part designed.

(Instructor: Note that the data in the table are such that one gear box can be designed to satisfy each group of five problems; in fact, the data have been taken from speed-reducer catalogs, so some manufacturer has already done this.) If this single gear box is an objective, all students in each group should use the same center distance and gear case and, if possible, the same shafts and bearings. Each student should be responsible for the solution of one problem but must cooperate with the group. This plan should result in an economic saving for limited production, even though some problems are overdesigned.

738–750. These numbers may be used for other problems.

Section 13

BEVEL GEARS

DESIGN PROBLEMS

Note to instructor: See also **761–776.**

751. Decide upon the pitch, face, N_g, material, and heat treatment of a pair of straight bevel gears to transmit continuously and indefinitely a uniform loading of 5 hp at 900 rpm of the pinion; reasonable operating temperature, high reliability; $m_g \approx 1.75$; $D_p \approx 3.333$ in. Pinion overhangs, gear is straddle mounted.

752. A pair of steel Zerol bevel gears is to transmit 25 hp at 600 rpm of the pinion; $m_g \approx 3$; let $N_p \approx 20$ teeth; highest reliability; the pinion is overhung, the gear straddle mounted. An electric motor drives a multicylinder pump. (a) Decide upon the pitch, face width, diameters, and steel (with treatment) for intermittent service. (b) The same as (a) except that indefinite life is desired.

753. Decide upon the pitch, face, and number of teeth for two spiral-bevel gears for a speed reducer. The input to the pinion is 20 hp at 1750 rpm; $m_g \approx 1.9$; pinion overhung, gear straddle mounted. It is hoped not to exceed a maximum D_p of $4\frac{3}{8}$-in.; steel gears with minimum 245 BHN on pinion and 210 BHN on gear. The gear is motor-driven, subject to miscellaneous drives involving moderate

shock; indefinite life against breakage and wear with high reliability. If the gears designed for the foregoing data are to be subjected to intermittent service only, how much power could they be expected to transmit?

754. The data are the same as in **753,** except that it has been decided to check the design for carburized teeth with the aim of reducing weight; delete $D_p \approx 4\frac{3}{8}$ in. and let the minimum $N_p \approx 20$ teeth.

CHECK PROBLEMS

755. A pair of straight-bevel gears are to transmit a smooth load of 45 hp at 500 rpm of the pinion; $m_g = 3$. A proposed design is $D_g = 15$ in., $b = 2\frac{3}{8}$ in., $P_d = 4$. Teeth are carburized AISI 8620, SOQT 450°F. The pinion overhangs, the gear is straddle-mounted. Would these gears be expected to perform with high reliability in continuous service? If not would you expect more than 1 failure in 100?

756. A gear catalog rates a pair of cast-iron, straight-bevel gears at 15.26 hp at 800 rpm of the 16-tooth pinion; $m_g = 3.5$, $b = 3$ in., $P_d = 3$; pinion over-

74

hangs, straddle-mounted gear. Assume the cast iron to be class 30. If the load is smooth, is this rating satisfactory, judging by the design approach of the *Text* for good reliability (a) when strength alone is considered, (b) when long continuous service is desired?

757. An 870-rpm motor drives a belt conveyor through bevel gears having 18 and 72 teeth; $P_d = 6$, $b = 1\frac{3}{4}$ in. Both gears are straddle-mounted. What horsepower may these gears transmit for an indefinite life with high reliability if both gears are (a) cast-iron, class 40; (b) AISI 5140, OQT 1000°F; (c) AISI 5140, OQT 1000°F, flame hardened; (d) AISI 8620, SOQT 450°?

758. A pair of straight-bevel gears transmits 15 hp at a pinion speed of 800 rpm; $P_d = 5$, $N_p = 20$, $N_g = 60$, $b = 2$ in. Both gears are made of AISI 4140 steel, OQT 800°F. What reliability factor is indicated for these gears for strength and for wear (a) for smooth loads, (b) for light shock load from the power source and heavy shock on the driven machine?

MISCELLANEOUS

759. In the differential shown, $N_A = 10$, $N_B = 47$ teeth, $P_d = 4.82$, $b = 1^{13}\!/_{32}$; the spiral angle $\psi = 39.511°$, left-hand on pinion A; $\phi_n = 20°$. The maximum brake engine torque is 2424 in-lb. at 2500 rpm of the engine. The low-gear reduction ahead of the differential is $m_\omega = 2.864$; in high gear, 1. Pinion A rotates counterclockwise. At maximum torque, compute the axial thrust on each gear (A,B) when the transmission is (a) in low gear with an efficiency of 93%, (b) in high gear with $e \approx 100\%$.

760. Two spiral bevels are used in an automotive differential similar to that shown; $N_A = 9$, $N_B = 41$, $P_d = 4.43$, $b = 1.25$ in.; $\psi = 45°$, left-hand on the pinion; $\phi = 20°$; in low gear, $m_g = 2.9$. The maximum torque on the pinion is 1980 in-lb. at 2500 rpm. Ring gear B is left (instead of right) of the center line in the illustration and A turns clockwise (from top). Compute the axial thrusts on A and B when the transmission is (a) in

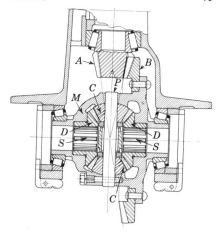

Problems 759, 760.

Pinion A turns with drive shaft. Ring gear B is riveted to part M, which carries pin P, on which rotate the planets C. Into the splines S in the hubs of gears D fit the rear axles. Hence, gears D and the rear axles turn together. When the vehicle moves straight ahead, part M, gears C and D, and axles all turn as a unit. When the vehicle moves in a curve, the outside wheel turns faster than the inside one, so that the gear D on the outside turns faster than the other D; planets C move in relation to D to accommodate the different speeds of the wheels.

low gear, efficiency $e = 93\%$, (b) in high gear, $e \approx 100\%$.

761 776. Two bevel gears, one straddle mounted and one overhanging,

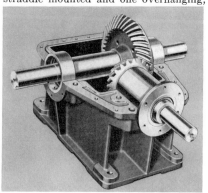

Problems 761-776.

Prob. No.	Approx. m_ω	Pinion Rpm	hp
761	2.9	600	40
762	4.6	900	60
763	4.4	900	50
764	4.5	1200	70
765	4.1	900	70
766	3.5	1200	50
767	5.3	1200	95
768	6.1	1800	130
769	4.2	1800	100
770	8.3	1800	80
771	4.6	1800	140
772	4.9	1200	80
773	7.2	1200	60
774	5.5	600	45
775	6.7	600	30
776	6.5	1800	120

are used in a reducer similar to that shown. The input is from an electric motor, to machines that impose moderate shock. The service is continuous and the gears should have indefinite life with the highest reliability. Let $N_p \approx 20$ teeth. Design first for spiral bevels. (a) Decide upon the pitch, face width, and steel (with treatment). (b) If a customer wishes to purchase your gear box for intermittent service, what maximum hp would you recommend for the same speed? (c) At the option of the instructor: design the shafts, keys, bearings (or select rolling types), and decide on the other dimensional details of the gears and case. See Fig. 15.18, *Text*, and manufacturers' catalogs for help on details. Sketch each part on a separate sheet of paper that includes complete information.

777–790. These numbers may be used for other problems.

Section 14

WORM GEARS

DESIGN PROBLEMS

Note to instructor: see also **806–830**.

791. (a) Determine a standard circular pitch and face width for a worm-gear drive with an input of 2 hp at 1200 rpm of the triple-threaded worm; the 1.58-in. (D_w) worm is steel with a minimum BHN = 250; gear is manganese bronze (Table AT 3); $m_\omega = 12$. Consider wear and strength only. Use a ϕ_n to match the lead angle λ. (See § 16.13, *Text.*) (b) Compute the efficiency.

792. A high-efficiency worm-gear speed reducer is desired, to accept 20 hp from a 1750-rpm motor. The diameter D_w of the integral worm has been estimated to be $1\frac{7}{8}$ in.; the next computations are to be for a steel worm with a minimum BHN = 250; phosphor-bronze gear (Table AT 3); $m_\omega = 11$. Probably, the worm should not have less than 4 threads. (a) Considering wear and strength only (§ 16.13), decide upon a pitch and face width that satisfies these requirements (§ 16.11, *Text*); specifying the pressure angle, diameters, and center distance. How does D_w used compare with that from equation (m), § 16.11, *Text?* What addendum and dedendum are recommended by Dudley? Compute a face length for the worm. (b) Compute

the efficiency. What do you recommend as the next trial for a "better" reducer?

793. The input to a worm-gear set is to be 25 hp at 600 rpm of the worm with $m_\omega = 20$. The hardened-steel worm is to be the shell type with a diameter approximately as given in § 16.11, *Text,* and a minimum of 4 threads; the gear is to be chilled phosphor bronze (Table AT 3). (a) Considering wear and strength only, determine suitable values of the pitch and face width. Let ϕ_n be appropriate to the value of λ. (b) Compute the efficiency. (c) Estimate the radiating area of the case and compute the temperature rise of lubricant. Is special cooling needed?

794. A 50-hp motor turning at 1750 rpm is to deliver its power to a worm-gear reducer, whose velocity ratio is to be 20. The shell-type worm is to be made of high-test cast iron; since a reasonably good efficiency is desired, use at least 4 threads; manganese-bronze gear (Table AT 3). (a) Decide upon D_w and ϕ_n, and determine suitable values of the pitch and face width. Compute (b) the efficiency, (c) the temperature rise of the lubricant. Estimate the radiating area of the case. Is special cooling needed?

CHECK PROBLEMS

795. A worm-gear speed reducer has a hardened-steel worm and a manganese-bronze gear (Table AT 3); triple-threaded worm with $P_c = 1.15278$ in., $D_w = 3.136$ in., $\phi_n = 25°$, $b = 2\frac{1}{4}$ in., $m_\omega = 12$, $n_w = 580$ rpm. The output is 16 hp. Compute (a) the dynamic load, (b) the endurance strength of the teeth and the indicated service factor on strength, (c) the limiting wear load (is it good for indefinitely continuous service?), (d) the efficiency and input hp, (e) the temperature rise of the oil (estimate case area as A_{min}, § 16.6). (f) Determine the tangential and radial components of the tooth load. (g) Is this drive self-locking?

796. For a worm-gear set, $P_c = 0.625$ in., $m_\omega = 8$, $D_w = 1.634$ in., $b = 1\frac{1}{8}$ in., $\phi_n = 25°$; 4-threaded worm turns at 1750 rpm. The worm is 250-BHN steel, and the gear, manganese bronze. Compute (a) the tooth's endurance strength, (b) the limiting wear load, (c) the safe horsepower output for continuous service; (d) (e), (f), (g) are the same as in **795.**

797. A worm-gear speed reducer has a hardened-steel worm and a phosphor-bronze gear. The lead angle of the 5-threaded worm $\lambda = 28° \ 57'$, $P_c = 1.2812$ in., $\phi_n = 25°$, $b = 2\frac{1}{2}$ in., $m_\omega = 8$; worm speed = 1750 rpm. The gear case is $35\frac{3}{8}$ in. high, 22 in. wide, 14 in. deep. Compute (a) the efficiency, (b) the limiting wear load, the strength load, and the corresponding safe input and output horsepowers. (c) The manufacturer rates this reducer at 53-hp input. Is this rating conservative or risky? (d) What is the calculated temperature rise of the oil with no special cooling? (e) The manufacturer specifies that for continuous service power should not exceed 36.5 hp if there is to be no artificial cooling and if Δt is to be less than 90°F. Make calculations and decide whether the vendor is on the safe side. (*Data courtesy of The Cleveland Worm Gear Co.*)

798. A worm-gear reducer has a hardened-steel, 4-threaded worm and a phosphor-bronze gear; $\lambda = 21.5°$, $P_c = 1.5163$ in., $\phi_n = 26°$, $b = 3\frac{1}{2}$ in., $m_\omega = 13$, worm speed = 1150 rpm. The gear case is 50 in. high, 30 in. wide, 20 in. deep.

Parts (a)–(e) are the same as in **797** except that in (c) the rated hp = 79.5, in (e) the limiting power without overheating is 52.5 hp. (*Data courtesy of The Cleveland Worm Gear Co.*).

HEATING

799. The input to a worm-gear reducer is 50.5 hp at 580 rpm of the 4-threaded worm. The gear case is $22 \times 31 \times 45$ in. in size; $\phi_n = 25°$, $P_c = 1.5$ in., $D_w = 4.432$ in., $f = 0.035$, room temperature = 80°F. Compute the steady-state temperature for average cooling.

800. The same as **799**, except that hp = 67.5 and $n = 1150$ rpm.

801. A hardened-steel, 4-threaded worm drives a bronze gear; $D_w = 1.875$ in., $D_g \approx 14$ in., $P_c = 1$ in., $\phi_n = 25°$, area of case ≈ 1500 sq. in., $v_r \approx 1037$ fpm; input = 20 hp at 1750 rpm of the worm; room temperature = 80°F. Compute the steady-state temperature of the lubricant for average ventilation.

802. The input to a 4-threaded worm is measured to be 20.8 hp; $P_c = 1$ in., $D_w = 2$ in., $\phi_n = 25°$. The area of the case is closely 1800 sq. in.; ambient temperature = 100°F; oil temperature = 180°F. Operation is at a steady thermal state. Compute the indicated coefficient of friction.

FORCE ANALYSIS

803. Make a force analysis of a worm and gear with the *gear* driving the worm. (a) Derive an equation for the efficiency of the drive. (b) Derive an equation that expresses the relationship between the lead angle, pressure angle, and coefficient of friction if the drive is self-locking. (c) The answer to (b) is $f \geqq \cos \phi_n \tan \lambda$. Show that when this condition of worm driving exists, the efficiency is less than 50%.

804. The input to a 4-threaded worm is 21 hp at 1750 rpm; $e = 90\%$, $D_w = 2\frac{1}{4}$ in., $D_g = 14$ in., $N_g = 44$, $\phi_n = 25°$. (a) From the horsepowers in and out, compute the tangential forces on the worm W_t and the gear F_t. (b) Using this value of F_t, compute W_t from equation (k), § 16.8, *Text*. (Check?) (c) Compute the separating force. (d) What is the end thrust on the worm shaft? On the gear shaft?

805. The input to a 2-threaded worm is 10.5 hp at 580 rpm; $e = 85\%$, $D_w = 3.104$ in., $D_g = 13.25$ in., $P_c = 1.04$ in., $\phi_n = 20°$. (a)–(d) The same as in **804.**

DESIGN PROJECTS

Instructor: the problem statement is arranged so that the length can be varied by assigning parts, say (a) and (b), as appropriate to your objectives.

806–830. A high-efficiency worm-gear reducer is to be designed in accordance with the tabulated specifications. Since high efficiency is desired, the worm should be integral with the shaft; also the thread should be hardened to a high BHN. Let the gear be phosphor bronze (Table AT 3). Use a normal pressure angle to match the lead angle. See § 16.13, *Text*. (a) Considering wear and strength only, decide upon a pitch and face width, specifying ϕ_n, D_w, D_g, C, and face length for worm. (If there seems to be trouble in getting a high lead angle, say about 30° or more, a tentative minimum worm shaft diameter may be computed from $T = 6000\pi D^3/16$. The root diameter of the thread should be greater than this D, say $\frac{1}{4}$ in. In the end of course, the shaft must be checked to see if it is safe). (b) Before making a final decision on P_c and b, compute the efficiency. (c) Then make new assumptions as may seem appropriate and redesign. (d) If the foregoing conclusions look good, use A_m, equation (g), § 16.6, as the expected case area and compute the temperature of the oil for an ambient 90°F. If this temperature rise exceeds 90°, determine the amount of heat to be removed in order to avoid overheating. (e) What are the tangential forces on worm and gear and the separating force? (f) At the instructor's option, design the shafts and bearings, and decide on all other dimensions, including details of housing the gears. Make sketches of the individual parts giving information for manufacture.

831–840. These numbers may be used for other problems.

Prob. No.	Rpm Worm	hp (input)	m_g
806	1750	57.5	13
807	1150	43.5	15⅓
808	870	31.0	20
809	720	25.0	23½
810	580	18.0	30
811	1750	65.5	15
812	1150	50.5	17⅓
813	870	37.5	21½
814	720	27.0	28½
815	580	22.0	31½
816	1750	83.5	14⅔
817	1150	67.0	18
818	870	50.0	22½
819	720	39.0	27
820	580	29.6	32½
821	1750	113	16⅓
822	1150	123	12¼
823	870	84.0	17⅔
824	720	66.5	21½
825	580	50.0	26½
826	1750	157	13¼
827	1150	146	12⅕
828	870	90.5	20
829	720	65.5	26½
830	580	53.5	30

Section 15

FLEXIBLE POWER-TRANSMITTING ELEMENTS

LEATHER BELTS

DESIGN PROBLEMS

841. A belt drive is to be designed for $F_1/F_2 = 3$, while transmitting 60 hp at 2700 rpm of the driver D_1; $m_\omega \approx 1.85$; use a medium double belt, cemented joint; a squirrel-cage, compensator-motor drive with mildly jerking loads; center distance is expected to be about twice the diameter of the larger pulley. (a) Choose suitable iron-pulley sizes and determine the belt width for a maximum permissible $s = 300$ psi. (b) How does this width compare with that obtained by the ALBA procedure? (c) Compute the maximum stress in the straight part of the ALBA belt. (d) If the belt in (a) stretches until the tight tension $F_1 = 525$ lb., what is F_1/F_2?

842. A 20-hp, 1750-rpm, slip-ring motor is to drive a ventilating fan at 330 rpm. The horizontal center distance must be about 8 to 9 ft. for clearance, and operation is continuous, 24 hr./day. (a) What driving-pulley size is needed for a speed recommended as about optimum in the *Text?* (b) Decide upon a pulley size (iron or steel) and belt thickness, and determine the belt width by the ALBA tables. (c) Compute the stress from the general belt equation assuming that the applicable coefficient of friction is that suggested by the *Text*. (d) Suppose the belt is installed with an initial tension $F_o = 70$ lb./in. (§ 17.10), compute F_1/F_2 and the stress on the tight side if the approximate relationship of the operating tensions and the initial tension is $F_1^{1/2} + F_2^{1/2} = 2F_o^{1/2}$.

843. A 100-hp squirrel-cage, line-starting electric motor is used to drive a Freon reciprocating compressor and turns at 1140 rpm; for the cast-iron motor pulley, $D_1 = 16$ in.; $D_2 = 53$ in., a flywheel; cemented joints; $C = 8$ ft. (a) Choose an appropriate belt thickness and determine the belt width by the ALBA tables. (b) Using the design stress of § 17.6, compute the coefficient of friction that would be needed. Is this value satisfactory? (c) Suppose that in the beginning, the initial tension was set so that the operating $F_1/F_2 = 2$. Compute the maximum stress in a straight part. (d) The approximate relation of the operating tensions and the initial tension F_o is $F_1^{1/2} + F_2^{1/2} = 2F_o^{1/2}$. For the condition in (c), compute F_o. Is it reasonable compared to Taylor's recommendation?

844. A 50-hp compensator-started motor running at 865 rpm drives a reciprocating compressor for a 40-ton refrig-

80

erating plant; flat leather belt, cemented joints. The diameter of the fiber driving pulley is 13 in., $D_2 = 70$ in., a cast-iron flywheel; $C = 6$ ft., 11 in. Because of space limitations, the belt is nearly vertical; the surroundings are quite moist. (a) Choose a belt thickness and determine the width by the ALBA tables. (b) Using recommendations in the *Text*, compute s from the general belt equation. (c) With this value of s, compute F_1 and F_1/F_2. (d) Approximately, $F_1^{1/2} + F_2^{1/2} = 2F_o^{1/2}$, where F_o is the initial tension. For the condition in (c), what should be the initial tension? Compare with Taylor, § 17.10. (e) Compute the belt length. (f) The data are from an actual drive. Do you have any recommendations for redesign on a more economical basis?

845. A 37.5-kva generator, running at 1200 rpm, is belt driven from a Diesel-engine flywheel 56 in. in diameter; $D_1 = 9.5$ in. (generator shaft); $C = 106$ in., horizontal; use machine wire-laced, medium double leather belt. (a) Determine the belt width by the ALBA tables. (b)–(f) as in **844.**

CHECK PROBLEMS

846. An exhaust fan in a wood shop is driven by a belt from a squirrel-cage motor that runs at 880 rpm, compensator started. A medium double leather belt, 10 in. wide, is used; $C = 54$ in.; $D_1 = 14$ in. (motor), $D_2 = 54$ in., both iron. (a) What horsepower, by ALBA tables, may this belt transmit? (b) For this power, compute the stress from the general belt equation. (c) For this stress, what is F_1/F_2? (d) If the belt has stretched until $s = 200$ psi on the tight side, what is F_1/F_2? (e) Compute the belt length.

847. A motor is driving a centrifugal compressor through a 6-in. heavy, single-ply leather belt in a dusty location. The 8-in. motor pulley turns 1750 rpm; $D_2 = 12$ in. (compressor shaft); $C = 5$ ft. The belt has been designed for a net belt pull of $F_1 - F_2 = 40$ lb./in. of width and $F_1/F_2 = 3$. Compute (a) the horsepower, (b) the stress in the tight side. (c) For this stress, what needed value of f is indicated by the general belt equation? (d) Considering the original data, what horsepower is obtained from the ALBA tables? Any remarks?

848. A 10-in. medium double leather belt, cemented joints, transmits 60 hp from a 9-in. paper pulley to a 15-in. pulley on a mine fan; dusty conditions. The compensator-started motor turns 1750 rpm; $C = 42$ in. This is an actual installation. (a) Determine the horsepower from the ALBA tables. (b) Using the general equation, determine the horsepower for this belt. (c) Estimate a service factor from Table 17.7 and apply it to the answer in (b). Does this result in better or worse agreement of (a) and (b)? What is your opinion as to the life of the belt?

MISCELLANEOUS

849. Let the coefficient of friction be constant. Find the speed at which a leather belt may transmit maximum power if the stress in the belt is (a) 400 psi, (b) 320 psi. (c) How do these speeds compare with those mentioned in § 17.9, *Text?* (d) Would the corresponding speeds for a rubber belt be larger or smaller? (HINT: Try the first derivative of the power with respect to velocity.)

850. A 40-in. pulley transmits power to a 20-in. pulley by means of a medium double leather belt, 20 in. wide; $C = 14$ ft., let $f = 0.3$. (a) What is the speed of the 40-in. pulley in order to stress the belt to 300 psi at zero power? (b) What maximum horsepower can be transmitted if the indicated stress in the belt is 300 psi? What is the speed of the belt when this power is transmitted? (See HINT in **849**).

AUTOMATIC TENSION DEVICES

851. An ammonia compressor is driven by a 100-hp synchronous motor that turns 1200 rpm; 12-in. paper motor pulley; 78-in. compressor pulley, cast iron; $C = 84$-in. A tension pulley is placed so that the angle of contact on the motor pulley is 193° and on the compressor pulley, 240°. A 12-in. medium double leather belt with a cemented joint is used. (a) What will be the tension in the tight side of the belt if the stress is 375 psi? (b) What will be the tension in the slack side? (c) What coefficient of friction is required on each pulley as indicated by the general equation? (d) What force must be exerted on the tension pulley to hold the belt tight, and what size do you recommend?

852. A 40-hp motor, weighing 1915 lb., runs at 685 rpm and is mounted on a pivoted base. In Fig. 17.11, *Text*, $e = 10$ in., $h = 19\frac{3}{16}$ in. The center of the $11\frac{1}{2}$-in. motor pulley is $11\frac{1}{2}$ in. lower than the center of the 60-in. driven pulley; $C = 48$ in. (a) With the aid of a graphical layout, find the tensions in the belt for maximum output of the motor if it is compensator started. What should be the width of the medium double leather belt if $s = 300$ psi? (c) What coefficient of friction is indicated by the general belt equation? (*Data courtesy of Rockwood Mfg. Co.*)

853. A 50-hp motor, weighing 1900 lb., is mounted on a pivoted base, turns 1140 rpm, and drives a reciprocating compressor; in Fig. 17.11, *Text*, $e = 8\frac{3}{4}$ in., $h = 17\frac{5}{16}$ in. The center of the 12-in. motor pulley is on the same level as the center of the 54-in. compressor pulley; $C = 40$ in. (a) With the aid of a graphical layout, find the tensions in the belt for maximum output of the motor if it is compensator started. (b) What will be the stress in the belt if it is a heavy double leather belt, 11 in. wide? (c) What coefficient of friction is indicated by the general belt equation? (*Data courtesy of Rockwood Mfg. Co.*)

RUBBER BELTS

854. A 5-ply rubber belt transmits 20 horsepower to drive a mine fan. An 8-in. motor pulley turns 1150 rpm; $D_2 = 36$ in., fan pulley; $C = 23$ ft. (a) Design a rubber belt to suit these conditions, using a net belt pull as recommended in § 17.15, *Text*. (b) Actually, a 9-in., 5-ply Goodrich high-flex rubber belt was used. What are the indications for a good life?

855. A 20-in., 10-ply rubber belt transmits power from a 300-hp motor, running at 650 rpm, to an ore crusher. The center distance between the 33-in. motor pulley and the 108-in. driven pulley is 18 ft. The motor and crusher are so located that the belt must operate at an angle of 75° with the horizontal. What is the overload capacity of this belt if the

rated capacity is as defined in § 17.15, *Text?*

V-BELTS

NOTE. *If manufacturer's catalogs are available, solve these problems from catalogs as well as from data in the Text.*

856. A centrifugal pump, running at 340 rpm, consuming 105 hp in 24-hr service, is to be driven by a 125-hp, 1180-rpm compensator-started motor; $C = 43$ to 49 in. Determine the details of a multiple V-belt drive for this installation. The B. F. Goodrich Company recommended six C195 V-belts with 14.4-in. and 50-in. sheaves; $C \approx 45.2$ in.

857. A 50-hp, 1160-rpm, AC split-phase motor is to be used to drive a reciprocating pump at a speed of 330 rpm. The pump is for 12-hr. service and normally requires 44 hp, but it is subjected to peak loads of 175% of full load; $C \approx 50$ in. Determine the details of a multiple V-belt drive for this application. The Dodge Manufacturing Corporation recommended a Dyna-V Drive consisting of six 5V1800 belts with 10.9-in. and 37.5-in. sheaves; $C \approx 50.2$ in.

858. A 200-hp, 600-rpm induction motor is to drive a jaw crusher at 125 rpm; starting load is heavy; operating with shock; intermittent service; $C = 113$ to 123 in. Recommend a multiple V-flat drive for this application. The B. F. Goodrich Company recommended eight D480 V-belts with a 26-in. sheave and a 120.175-in. pulley; $C \approx 116.3$ in.

859. A 150-hp, 700-rpm, slip-ring induction motor is to drive a ball mill at 195 rpm; heavy starting load; intermittent seasonal service; outdoors. Determine all details for a V-flat drive. The B. F. Goodrich Company recommended eight D270 V-belts, 17.24-in. sheave, 61-in. pulley, $C \approx 69.7$ in.

860. A 30-hp, 1160-rpm, squirrel-cage motor is to be used to drive a fan. During the summer, the load is 29.3 hp at a fan speed of 280 rpm; during the winter, it is 24 hp at 238 rpm; $44 < C < 50$ in.; 20 hr./day operation with no overload. Decide upon the size and number of V-belts, sheave sizes, and belt length.

(*Data courtesy of The Worthington Corporation.*)

POWER CHAINS

NOTE. *If manufacturer's catalogs are available, solve these problems from catalogs as well as from data in the Text.*

861. A roller chain is to be used on a paving machine to transmit 30 hp from the 4-cylinder Diesel engine to a countershaft; engine speed 1000 rpm, countershaft speed 500 rpm. The center distance is fixed at 24 in. The chain will be subjected to intermittent overloads of 100%. (a) Determine the pitch and the number of chains required to transmit this power. (b) What is the length of the chain required? How much slack must be allowed in order to have a whole number of pitches? A chain drive with significant slack and subjected to impulsive loading should have an idler sprocket against the slack strand. If it were possible to change the speed ratio slightly, it might be possible to have a chain with no appreciable slack. (c) How much is the bearing pressure between the roller and pin?

862. A conveyor is driven by a 2-hp high-starting-torque electric motor through a flexible coupling to a worm-gear speed reducer, whose $m_\omega = 35$, and then via a roller chain to the conveyor shaft that is to turn about 12 rpm; motor rpm is 1750. Operation is smooth, 8 hr./day. (a) Decide upon suitable sprocket sizes, center distance, and chain pitch. Compute (b) the length of chain, (c) the bearing pressure between the roller and pin. The Morse Chain Company recommended 15- and 60-tooth sprockets, 1-in. pitch, $C = 24$ in., $L = 88$ pitches.

863. A roller chain is to transmit 5 hp from a gearmotor to a wood-working machine, with moderate shock. The 1-in. output shaft of the gearmotor turns $n = 500$ rpm. The $1\frac{1}{4}$-in. driven shaft turns 250 rpm; $C \approx 16$ in. (a) Determine the size of sprockets and pitch of chain that may be used. If a catalog is available, be sure maximum bore of sprocket is sufficient to fit the shafts. (b) Compute the center distance and length of chain. (c) What method should be used to supply oil to the chain? (d) If a catalog is

available, design also for an inverted tooth chain.

864. A roller chain is to transmit 20 hp from a split-phase motor, turning 570 rpm, to a reciprocating pump, turning at 200 rpm; 24 hr./day service. (a) Decide upon the tooth numbers for the sprockets, the pitch and width of chain, and center distance. Consider both single and multiple strands. Compute (b) the chain length, (c) the bearing pressure between the roller and pin, (d) the factor of safety against fatigue failure (Table 17.8), with the chain pull as the force on the chain. (e) If a catalog is available, design also an inverted-tooth chain drive.

865. A $\frac{5}{8}$-in. roller chain is used on a hoist to lift a 500-lb. load through 14 ft. in 24 sec. at constant velocity. If the load on the chain is doubled during the speed-up period, compute the factor of safety (a) based on the chain's ultimate strength, (b) based on its fatigue strength. (c) At the given speed, what is the chain's rated capacity ($N_s = 20$ teeth) in hp? Compare with the power needed at the constant speed. Does it look as though the drive will have a "long" life?

WIRE ROPES

866. In a coal-mine hoist, the weight of the cage and load is 20 kips; the shaft is 400 ft. deep. The cage is accelerated from rest to 1600 fpm in 6 sec. A single 6 × 19, IPS, $1\frac{3}{4}$-in. rope is used, wound on an 8-ft. drum. (a) Include the inertia force but take the static view and compute the factor of safety with and without allowances for the bending load. (b) If $N = 1.35$ based on fatigue, what is the expected life? (c) Let the cage be at the bottom of the shaft and ignore the effect of the rope's weight. A load of 14 kips is gradually applied on the 6-kip cage. How much is the deflection of the cable due to the load and the additional energy absorbed? (d) For educational purposes and for a load of $0.2F_u$, compute the energy that this 400-ft. rope can absorb and compare it with that for a 400-ft., $1\frac{3}{4}$-in., as-rolled-1045 steel rod. Omit the weights of the rope and rod. What is the energy per pound of material in each case?

867. The same as **866**, except that $D_r = 1\frac{1}{2}$ in.

868. A hoist in a copper mine lifts ore a maximum of 2000 ft. The weight of car, cage, and ore per trip is 10 kips, accelerated in 6 sec. to 2000 fpm; drum diameter is 6 ft. Use a 6 × 19 plow-steel rope. Determine the size (a) for a life of 200,000 cycles and $N = 1.3$ on the basis of fatigue, (b) for $N = 5$ by equation (v), § 17.25, *Text*. (c) What is the expected life of the rope found in (b) for $N = 1.3$ on the basis of fatigue? (d) If a loaded car weighing 7 kips can be moved gradually onto the freely hanging cage, how much would the rope stretch? (e) What total energy is stored in the rope with full load at the bottom of the shaft? Neglect the rope's weight for this calculation. (f) Compute the pressure of the rope on the cast-iron drum. Is it reasonable?

869. For a mine hoist, the cage weighs 5900 lb., the cars 2100 lb., and the load of coal in the car 2800 lb.; one car load at a time on the hoist. The drum diameter is 5 ft., the maximum depth is 1500 ft. It takes 6 sec. to accelerate the loaded cage to 3285 fpm. Decide on a grade of wire and the kind and size of rope on the basis of (a) a life of 2×10^5 cycles and $N = 1.3$ against fatigue failure, (b) static considerations (but not omitting the inertia effect) and $N = 5$. (c) Make a final recommendation. (d) If the loaded car can be moved gradually onto the freely hanging cage, how much would the rope stretch? (e) What total energy has the rope absorbed, fully loaded at the bottom of the shaft? Neglect the rope's weight for this calculation. (f) Compute the pressure of the rope on the cast-iron drum. Is it all right?

870. The wire rope of a hoist with a short lift handles a total maximum load of 14 kips each trip. It is estimated that the maximum number of trips per week will be 1000. The rope is 6 × 37, IPS, 1⅜ in. in diameter, with steel core. (a) On the basis of $N = 1$ for fatigue, what size drum should be used for a 6-yr. life? (b) Because of space limitations, the actual size used was a 2.5-ft. drum. What is the factor of safety on a static basis? What life can be expected $(N = 1)$?

871. A wire rope passes about a driving sheave making an angle of con-tact of 540°, as shown. A counterweight of 3220 lb. is suspended from one side and the acceleration is 4 fps². (a) If $f = 0.1$, what load may be raised without slipping on the rope? (b) If the sheave is rubber lined and the rope is dry, what load may be raised without slipping? (c) Neglecting the stress caused by bending about the sheave, find the size of 6 × 19 MPS rope required for $N = 6$ and for the load found in (a). (d) Compute the diameter of the sheave for indefinite life with say $N = 1.1$ on fatigue. What changes could be made in the solution to allow the use of a smaller sheave?

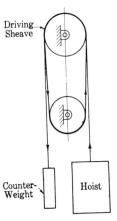

Problems 871-874.

872. A traction elevator with a total weight of 8 kips has an acceleration of 3 fps²; the 6 cables pass over the upper sheave twice, the lower one once, as shown. Compute the minimum weight of counterweight to prevent slipping on the driving sheave if it is (a) iron with a greasy rope, (b) iron with a dry rope, (c) rubber lined with a greasy rope. (d) Using MPS and the combination in (a), decide upon a rope and sheave size that will have indefinite life $(N = 1$ will do). (e) Compute the factor of safety defined in the *Text*. (f) If it were decided that 5×10^5 bending cycles would be enough life, would there be a significant difference in the results?

873. A 5000-lb. elevator with a traction drive is supported by 6 wire ropes, each passing over the driving sheave twice, the idler once, as shown.

Maximum values are 4500-lb. load, 4 fps² acceleration during stopping. The brake is applied to a drum on the motor shaft, so that the entire decelerating force comes on the cables, whose maximum length will be 120 ft. (a) Using the desirable D_s in terms of D_r, decide on the diameter and type of wire rope. (b) For this rope and $N = 1.05$, compute the sheave diameter that would be needed for indefinite life. (c) Compute the factor of safety defined in the *Text* for the result in (b). (d) Determine the minimum counterweight to prevent slipping with a dry rope on an iron sheave. (e) Compute the probable life of the rope on the sheave found in (a) and recommend a final choice.

874. A traction elevator has a maximum deceleration of 8.05 fps² when being braked on the downward motion with a total load of 10 kips. There are 5 cables that pass twice over the driving sheave. The counterweight weighs 8 kips. (a) Compute the minimum coefficient of friction needed between ropes and sheaves for no slipping. Is a special sheave surface needed? (b) What size 6 × 19 mild-plow-steel rope should be used for $N = 4$, including the bending effect? (Static approach.) (c) What is the estimated life of these ropes $(N = 1)$?

875–880. These numbers may be used for other problems.

Section 16

BRAKES AND CLUTCHES

ENERGY TO BRAKES

881. A motor operates a hoist through a pair of spur gears, with a velocity ratio of 4. The drum on which the cable wraps is on the same shaft as the gear, and the torque caused by the weight of the load and hoist is 12,000 ft-lb. The pinion is on the motor shaft. Consider first on which shaft to mount the brake drum; in the process make trial calculations, and try to think of pros and cons. Make a decision and determine the size of a drum that will not have a temperature rise greater than $\Delta t = 150°F$ when a 4000-lb. load moves down 200 ft. at a constant speed. Include a calculation for the fhp/sq. in. of the drum's surface.

882. A 3500-lb. automobile, moving on level ground at 60 mph, is to be stopped in a distance of 260 ft. Tire diameter is 30 in.; all frictional energy except for the brake is to be neglected. (a) What total average braking torque must be applied? (b) What must be the minimum coefficient of friction between the tires and the road in order for the wheels not to skid if it is assumed that weight is equally distributed among the four wheels (not true)? (c) If the frictional energy is momentarily stored in 50 lb. of cast-iron brake drums, what is the average temperature rise of the drums?

883. The same as **882**, except that the car is moving down a 5% grade.

884. An overhead traveling crane weighs 160,000 lb. with its load and runs 253 fpm. It is driven by a 25-hp motor operating at 1750 rpm. The speed reduction from the motor to the 18-in. wheels is 32 to 1. Frictional energy other than at the brake is negligible. (a) How much energy must be absorbed by the brake to stop this crane in a distance of 18 ft.? (b) Determine the constant average braking torque that must be exerted on the motor shaft. (c) If all the energy is absorbed by the rim of the cast iron brake drum, which is 8 in. in diameter, $\frac{1}{2}$ in. thick, with a $3\frac{1}{4}$-in. face, what will be its temperature rise? (d) Compute the average rate at which the energy is absorbed during the first second (fhp). Is it reasonable?

885. The diagrammatic hoist shown with its load weighs 6000 lb. The drum weighs 8000 lb., has a radius of gyration $k = 1.8$ ft.; $D = 4$ ft. A brake on the drum shaft brings the hoist to rest in 10 ft. from $v_s = 8$ fps (down). Only the brake frictional energy is significant, and it can be reasonably assumed that the acceleration is constant. (a) From the frictional energy, compute the average braking torque. (b) If the average fhp/ sq. in. is limited to 0.15 during the first second, what brake contact area is needed?

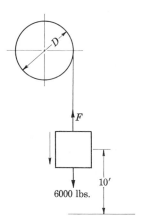

Problems 885, 886.

886. The same as **885,** except that $v_s = 12$ fps.

887. The same as **885,** except that a traction drive, arranged as shown, is used; the counterweight weighs 4000 lb. The ropes pass twice about the driving sheave; the brake drum is on this same shaft.

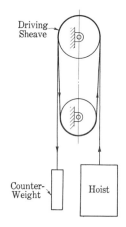

Problem 887.

SINGLE-SHOE BRAKES

888. For the single-shoe, short-block brake shown (solid lines) derive the expressions for brake torque for (a) clock-

wise rotation, (b) counterclockwise rotation. (c) In which direction of rotation does the brake have self-actuating properties? If $f = 0.25$, for what proportions of e and c would the brake be self-actuating?

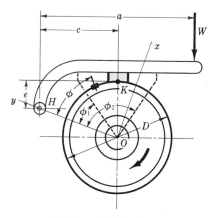

Problems 888–891, 893.

889. The same as **888,** except that the wheel and brake shoe are grooved, 2θ degrees between the sides of the grooves (as in a sheave, Fig. 17.38, *Text*).

890. Consider the single-shoe, short-block brake shown (solid lines) with the drum rotating clockwise; let e be positive measured downward and $D = 1.6c$. (a) Plot the mechanical advantage MA (ordinate) against f values of 0.1, 0.2, 0.3, 0.4, 0.5 (abscissa) when e/c has values 2, 0.5, 0, −0.5, −1. (b) If f may vary from 0.3 to 0.4, which proportions give the more nearly constant brake response? Are proportions good? (c) What proportions are best if braking is needed for both directions of rotation?

891. A single-block brake has the dimensions: cast-iron wheel of $D = 15$ in., $a = 32\frac{1}{2}$ in., $c = 9\frac{3}{8}$ in., $e = 4\frac{11}{16}$ in., width of contact surface = 2 in. The brake block lined with molded asbestos, subtends 80°, symmetrical about the center line; it is permitted to absorb energy at the rate of 0.4 hp/in.^{2}; $n = 200$ rpm. Assume that p is constant, that F and N act at K, and compute (a) pv_m and the approximate braking torque, (b) the force W to produce this torque, (c) the mechanical advantage, (d) the temperature rise of the $\frac{3}{8}$-in.-thick rim, if it absorbs all the energy with operation as specified, in 1 min. (e) How long could this brake be so applied for $\Delta t = 400°F$? See **893**.

892. For a single-block brake, as shown, $a = 26$ in., $c = 7\frac{1}{2}$ in., $e = 3.75$ in., $D = 15$ in., drum contact width $b = 3\frac{1}{2}$ in. The molded asbestos lining subtends $\theta = 60°$, symmetrical about the vertical axis; force $W = 300$ lb.; $n = 600$ rpm. Assume that p is constant, that F and N act at K, and compute (a) pv_m and the braking torque, (b) the energy rate in fhp/in.² of contact surface, (c) the mechanical advantage, (d) the temperature of the $\frac{3}{8}$-in.-thick rim, if it absorbs all the energy with the operation as specified in 1 min. (e) How long could this brake be so applied for $\Delta t_{rim} = 400°F$? See **894**.

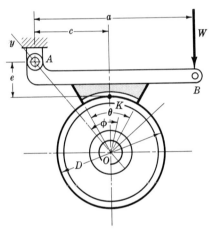

Problems 892, 894.

LONG-SHOE BRAKES

FIXED SHOES

893. The brake is as described in **891** and is to absorb energy at the same rate but the pressure varies as $p = P \sin \theta$. Derive the equations needed and compute (a) the maximum pressure, (b) the moment $M_{F/H}$ of F about H, (c) the moment $M_{N/H}$ of N about H, (d) the force W, (e) the braking torque, (f) the x and y components of the force at H.

894. The brake is as described in **892**, but the pressure varies as $p = P \sin \phi$. Assume the direction of rotation

for which a given W produces the greater T_f, derive the equations needed, and compute (a) the maximum pressure, (b) the moment of F about A, (c) the moment of N about A, (d) the braking torque, (e) the x and y components of the force at A.

895. (a) For the brake shown, assume $p = P \cos \alpha$ and the direction of rotation for which a given force W results in the greater braking torque, and derive equations for T_f in terms of W, f, and the dimensions of the brake. (b) Under what circumstances will the brake be self-acting? (c) Determine the magnitude and location of the resultant forces N and F.

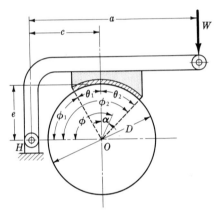

Problems 895, 896.

896. For the brake shown with $\theta_1 \neq \theta_2$, assume that the direction of rotation is such that a given W results in the greater braking torque and that $p = P \sin \phi$. (a) Derive equations in terms of θ_1 and θ_2 for the braking torque, for the moment $M_{F/H}$, and for $M_{N/H}$. (b) Reduce the foregoing equations for the condition $\theta_1 = \theta_2$. (c) Now suppose that θ, taken as $\theta = \theta_1 + \theta_2$, is small enough that $\sin \theta \approx \theta$, $\cos \theta \approx 1$, $\theta_1 = \theta_2 = \theta/2$. What are the resulting equations?

897. The brake shown is lined with woven asbestos; the cast-iron wheel is turning at 60 rpm CC; width of contact surface is 4 in. A force $W = 1300$ lb. is applied via a linkage system not shown; $\theta = 90°$. Let $p = P \sin \phi$. (a) With the brake lever as a free body, take moments about the pivot J and determine the maximum pressure and compare with permissible values. Compute (b) the braking torque, (c) the frictional energy in fhp. (d) Compute the normal force N, the average pressure on the projected area, and decide if the brake application can safely be continuous.

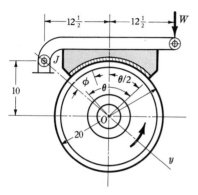

Problem 897.

PIVOTED-SHOE BRAKES

898. In the brake shown, the shoe is lined with flexible woven asbestos, and pivoted at point K in the lever; face width is 4 in.; $\theta = 90°$. The cast-iron wheel turns 60 rpm CL; let the maximum pressure be the value recommended in Table AT 29. On the assumption that K will be closely at the center of pressure, as planned, compute (a) the brake torque, (b) the magnitude of force W, (c) the rate at which frictional energy grows, (d) the time of an application if it is assumed

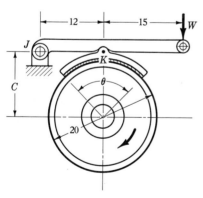

Problem 898.

that all this energy is stored in the 1-in. thick rim with $\Delta t_{rim} = 350°F$, (e) the average pressure on projected area. May this brake be applied for a "long time" without damage? (f) What would change for CC rotation?

899. The pivoted-shoe brake shown is rated at 450 ft-lb. of torque; $\theta = 90°$; contact width is 6.25 in.; cast-iron wheel turns at 600 rpm; assume a symmetric sinusoidal distribution of pressure. (a) Locate the center of pressure and compare with the location of K. Compute (b) the maximum pressure and compare with allowable value, (c) the value of force W, (d) the reaction at the pin H, (e) the average pressure and pv_m, and decide whether or not the application could be continuous at the rated torque. (f) Compute the frictional work from $T\omega$ and estimate the time it will take for the rim temperature to reach 450°F (ambient, 100°F).

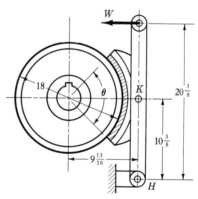

Problem 899.

TWO-SHOE BRAKES

PIVOTED SHOES

900. The double-block brake shown is to be used on a crane; the force W is applied by a spring, and the brake is released by a magnet (not shown); $\theta = 90°$; contact width = 2.5 in. Assume that the shoes are pivoted at the center of pressure. The maximum pressure is the permissible value of Table AT 29. Compute (a) the braking torque, (b) the force W, (c) the rate of growth of frictional energy at 870 rpm, (d) the time it would take to raise the temperature of the

0.5-in.-thick rim by $\Delta t = 300°F$ (usual assumption of energy storage), (e) pv_m. (f) Where should the pivot center be for the calculations to apply strictly?

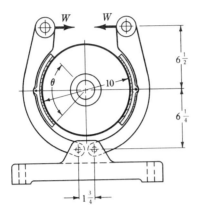

Problem 900.

901. A pivoted-shoe brake, rated at 900 ft-lb. torque, is shown. There are 180 sq. in. of braking surface; woven asbestos lining; 600 rpm of the wheel; 90° arc of brake contact on each shoe. The effect of spring A is negligible. (a) Is the pin for the shoe located at the center of pressure? (b) How does the maximum pressure compare with that in Table AT 29? (c) What load W produces the rated torque? (d) At what rate is energy absorbed? Express in horsepower. Is it likely that this brake can operate continuously without overheating? (e) Does the direction of rotation affect the effectiveness of this brake?

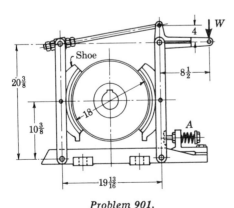

Problem 901.

902. Refer to the diagrammatic representation of the brake of Fig. 18.2, *Text*, and let the dimensions be: $a = b = m = t = 4\%_{16}$, $c = 14$, $D = 15$, $h = 9\%_{16}$ in., and the contact width is 4 in.; arc of contact $= 90°$; lining is asbestos in resin binder, wheel rotation of 100 rpm CC; applied load $W = 2000$ lb. (a) Locate the center of pressure for a symmetric sinusoidal pressure distribution and compare with the actual pin centers. Assume that this relationship is close enough for approximate results and compute (b) the dimensions k and e if the braking force on each shoe is to be the same, (c) the normal force and the maximum pressure, (d) the braking torque, (e) pv_m. Would more-or-less continuous application be reasonable?

FIXED SHOES

903. A double-block brake has certain dimensions as shown. Shoes are lined with woven asbestos; cast-iron wheel turns 60 rpm; applied force $W = 70$ lb. For each direction of rotation, compute (a) the braking torque, (b) the rate of generating frictional energy (fhp). (c) If the maximum pressure is to be $P = 50$ psi (Table AT 29), what contact width should be used? (d) With this width, compute pv_m and decide whether or not the applications must be intermittent.

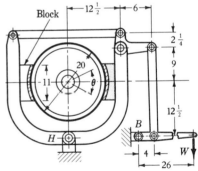

Problems 903, 904.

904. If the brake shown has a torque rating of 7000 lb-in. for counterclockwise rotation, what braking torque would it exert for clockwise rotation, force W the same?

905. A double-block brake is shown for which $\theta = 90°$, $b = 5$ in., $n = 300$

rpm, rim thickness $= \frac{3}{4}$ in., and $W = 400$ lb. The shoes are lined with asbestos in resin binder. Determine the frictional torque for (a) clockwise rotation, (b) counterclockwise rotation. (c) How much energy is absorbed by the brake? Express in horsepower. (d) Will the brake operate continuously without danger of overheating? How long for a $\Delta t_{\text{rim}} = 300°F$? How does pv_m compare with *Text* values?

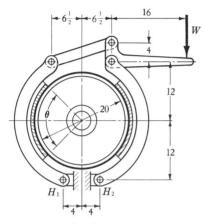

Problem 905.

906. The double-block brake for a crane has the dimensions: $a = 14.3$, $b = 2.37$, $c = 0.83$, $D = 10$, $e = 11.05$,

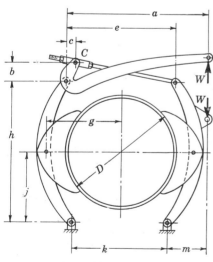

Problem 906.

$g = 7.1$, $h = 12$, $j = 6.6$, $k = 10.55$, $m = 3.5$ in., the width of shoes is 4 in., and the subtended angle is $\theta = 90°$; woven asbestos lining. Its rated braking torque is 200 ft-lb. The shoes contact the arms in such a manner that they are virtually fixed to the arms. What force W must be exerted by a hydraulic cylinder to develop the rated torque for (a) counterclockwise rotation, (b) clockwise rotation? Is the torque materially affected by the direction of rotation? (c) Compute the maximum pressure and compare with that in Table AT 29. (*Data courtesy of Wagner Electric Corporation.*)

907. The sketch shows a two-shoe magnetic brake that has a maximum torque rating of 200 ft-lb. The shoes are adjusted on installation to fit the position of the brake drum but are otherwise "fixed"; face width $b = 3\frac{3}{4}$ in. The brake is actuated by the spring force W, and is released when a magnet is energized. (The levers on the right are a diagrammatic representation.) The wheel is ductile iron and the brake lining is compressed asbestos in rubber binder. Determine the spring force W required for (a) clockwise rotation, (b) counterclockwise rotation. Does the direction of rotation make a material difference in the brake's effectiveness? (c) Determine the maximum pressure on each shoe for both directions of rotation. (*Data courtesy of Cutler-Hammer, Inc.*)

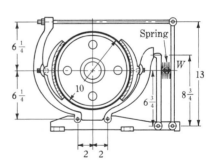

Problem 907.

INTERNAL-SHOE BRAKES

908. Assuming that the distribution of pressure on the internal shoe shown is given by $p = P \sin \phi$, show that the moments $M_{N/B}$, $M_{F/B}$, and $T_{F/O}$, of N

with respect to B and of F with respect to B and to O are (b = face width)

$$M_{N/B} = (Pbar/2)[\theta - (\sin 2\phi_2 - \sin 2\phi_1)/2],$$

$$M_{F/B} = fPbr[r(\cos \phi_1 - \cos \phi_2) - a(\sin^2 \phi_2 - \sin^2 \phi_1)/2],$$

$$T_{F/O} = fPbr^2(\cos \phi_1 - \cos \phi_2).$$

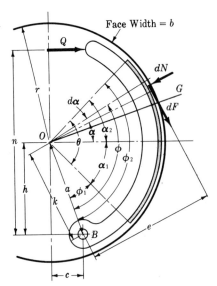

Face Width = b

Problems 908-910.

909. The same as **908,** except that a pressure distribution of $p = P \cos \alpha$ is assumed.

$$M_{N/B} = Pbr[h(2\theta + \sin 2\alpha_2 + \sin 2\alpha_1)/4 + c(\sin^2 \alpha_2 - \sin^2 \alpha_1)/2],$$

$$M_{F/B} = fPbr[r(\sin \alpha_2 + \sin \alpha_1) + h(\sin^2 \alpha_2 - \sin^2 \alpha_1)/2 - c(2\theta + \sin 2\alpha_2 + \sin 2\alpha_1)/4],$$

$$M_{F/O} = fPbr^2(\sin \alpha_2 + \sin \alpha_1).$$

910. The same as **909,** except that α is to be measured from OG, a perpendicular to OB; limits from $-\alpha_1$ to $+\alpha_2$.

911. The following dimensions apply to a two-shoe truck brake somewhat as shown: face $b = 5$, $r = 8$, $h = 5.1$, $c = 2.6$, $w = u = 6.4$ in., $\theta = 110°$, $\phi_1 = 15°$. Lining is asbestos in rubber compound. For a maximum pressure on each shoe of 100 psi, determine the force Q, and the braking torque for (a) clockwise rotation, (b) counterclockwise rota-

tion. See **908.** (*Data courtesy of Wagner Electric Corporation.*)

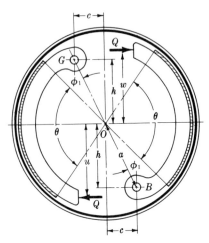

Problems 911, 912.

912. The same as **911** except that the equations of **909** are to be used.

913. The data are the same as **911,** but the shoe arrangement is as shown for this problem. For a maximum pressure on the shoes of 100 psi, determine the force Q and $T_{F/O}$ for (a) CL rotation, (b) CC rotation. See **908.**

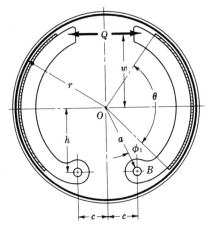

Problem 913.

914. A double-shoe internal brake is actuated by an involute cam as shown, where Q_R is the force on the right shoe at a radius w_R and Q_L is the force on the left shoe at a radius w_L. The pressure of each shoe is proportional to the rotation of the shoe about B which is inversely proportional to w; therefore, the ratio of the maximum pressures is $P_L/P_R = w_R/w_L$. The dimensions are: face width $b = 4$, $r = 6$, $h = 4\frac{9}{16}$, $c = 1\frac{1}{8}$, $w_L = 9\frac{5}{16}$, $w_R = 8\frac{5}{16}$ in.; for each shoe, $\theta = 120°$, $\phi_1 = 30°$. The lining is asbestos in rubber compound. Determine the braking torque and forces Q_R and Q_L for the maximum permissible pressure for (a) clockwise rotation, (b) counterclockwise rotation.

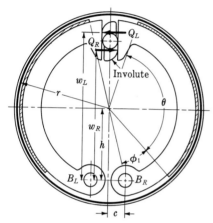

Problem 914.

BAND BRAKES

915. The steel band for the brake shown is lined with flexible asbestos and it is expected that the permissible pressure of Table AT 29 is satisfactory; $\theta = 245°$, $a = 20$ in., $m = 3\frac{1}{2}$ in., $D = 18$ in., and face width $b = 4$ in.; rotation CL. The cast-iron wheel turns

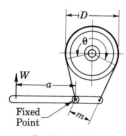

Problem 915.

200 rpm. Set up suitable equations, use the average f given and compute (a) the force on each end of the band, (b) the brake torque and fhp. (c) Determine the mechanical advantage for the limit values of f in Table AT 29 and its percentage variation from that for the average f. (d) Investigate the overheating problem, using relevant information given in the *Text*.

916. (a) For the band brake shown, derive the expressions for the braking torque in terms of W, etc., for CL rotation and for CC rotation, and specify the ratio of c/b for equal effectiveness in both directions of rotation. Are there any proportions of b and c as shown that would result in the brake being self locking? (b) When $\theta = 270°$, $a = 16$ in., $b = c = 3$ in., and $D = 12$ in., it was found that a force $W = 50$ lb. produced a frictional torque of 1000 in-lb. Compute the coefficient of friction.

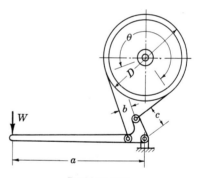

Problem 916.

917. (a) For the brake shown, assume the proper direction of rotation of the cast-iron wheel for differential action and derive expressions for the braking force and braking torque. (b) Let $D = 14$ in., $n = 1\frac{3}{4}$ in., $m = 4$ in., $\theta = 235°$, and assume the band to be lined with woven asbestos. Is there a chance that this brake will be self-acting? If true, will it always be for the range of values of f given in Table AT 29? (c) The ratio of n/m should exceed what value in order for the brake to be self-locking? (d) If the direction of rotation of the wheel is opposite to that taken in (a), what is the braking torque with a force $W = 10$ lb. at $a = 8$ in.? (e) Suppose the brake is used as a stop to prevent reverse motion on a hoist. What is the frictional horsepower for the forward motion if the wheel turns 63 rpm?

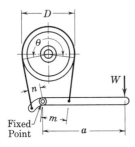

Problems 917, 918.

918. A differential band brake similar to that shown and lined with woven asbestos, has the dimensions: $D = 18$ in., $n = 2$ in., $m = 12$ in., $\theta = 195°$. (a) Is there a chance that this brake will be self-acting? (b) If $W = 30$ lb. and $a = 26$ in., compute the maximum braking torque and the corresponding mechanical advantage. (c) What is the ratio of the braking torque for CL rotation to the braking torque for CC rotation? (d) A $\frac{1}{16}$-in.-thick steel band, SAE 1020 as rolled, carries the asbestos lining. What should be its width for a factor of safety of 8, based on the ultimate stress? What should be the face width if the average pressure is 50 psi?

919. A differential band brake is to be designed to absorb 10 fhp at 250 rpm. (a) Compute the maximum and minimum diameters from both equations (z) and (a), p. 495, *Text*. Decide on a size. (b) The band is to be lined with woven asbestos. The Rasmussen recommendation (§ 18.4) will help in deciding on the face width. Also check the permissible pressure in Table AT 29. Choose dimensions of the lever, its location and shape and the corresponding θ. Be sure the brake is not self locking. What is the percentage variation of the mechanical advantage from the minimum value (f_{min}) for the f limits in Table AT 29?

DISK CLUTCHES

920. An automobile engine develops its maximum brake torque at 2800 rpm when the bhp = 200. A design value of $f = 0.25$ is expected to be reasonable for the asbestos facing and it is desired that the mean diameter not exceed 8.5 in.; permissible pressure is 35 psi. Designing for a single plate clutch, Fig. 18.10, *Text*, determine the outer and inner diameters of the disk.

921. An automobile engine can develop a maximum brake torque of 2448 in-lb. Which of the following plate clutches, which make up a manufacturer's standard "line," should be chosen for this car? Facing sizes: (a) $D_o = 8\frac{7}{8}$, $D_i = 6\frac{1}{8}$ in., (b) $D_o = 10$, $D_i = 6\frac{1}{8}$ in., (c) $D_o = 11\frac{1}{16}$, $D_i = 6\frac{1}{8}$ in. In each case, assume $f = 0.3$. The unit pressures are (a) 34 psi, (b) 30 psi, and (c) 26.2 psi.

922. A single-disk clutch for an industrial application, similar to that in Fig. 18.11, *Text*, except that there are two disks attached to one shaft and one attached to the other. The clutch is rated at 50 hp at 500 rpm. The asbestos-in-resin-binder facing has a $D_o = 8\frac{1}{2}$ in. and $D_i = 4\frac{3}{4}$ in. What must be the axial force and average pressure? How does this pressure compare with that recommended by Table AT 29?

923. A multiple-disk clutch similar to Fig. 18.11, *Text*, is rated at 22 hp at 100 rpm. The outside and inside diameters of the disks are 14 and $7\frac{1}{2}$ in., respectively. If $f = 0.25$, find (a) the axial force required to transmit the rated load, and (b) the unit pressure between the disks.

924. A multiple-disk clutch for a machine tool operation has 4 phosphor-bronze driving disks and 5 hardened-steel driven disks. This clutch is rated at 5.8 hp at 100 rpm when operated dry. The outside and inside diameters of the disks are $5\frac{1}{2}$ and $4\frac{3}{16}$ in., respectively. (a) If the pressure between the disks is that recommended for metal on metal in Table AT 29, what coefficient of friction is required to transmit the rated power? (b) What power may be transmitted for f and p as recommended in Table AT 29?

925. A multiple-disk clutch with three disks on one shaft and two on the other, similar to that in Fig. 18.11, *Text*, is rated at 53 hp at 500 rpm. (a) What is the largest value of D_i if f and p are given by Table AT 29 for asbestos in resin binder and $D_o = 10.5$ in. (b) For the diameter used of $D_i = 7$ in., what is the required axial force and the average pressure?

MISCELLANEOUS CLUTCHES AND BRAKES

926. For the cone brake shown, find an expression for the braking torque for a given applied force W on the bell crank. Consider the force F', Fig. 18.12, *Text*, in obtaining the expression.

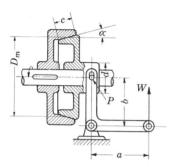

Problems 926-928.

927. For the cone brake similar to that shown, certain dimensions are: $D_m = 15$ in., $c = 2\frac{1}{2}$ in., $\alpha = 12°$, $b = 9$ in., and $a = 20$ in. The contact surfaces are metal and asbestos. (a) For an applied force $W = 80$ lb., what brak-ing torque may be expected of this brake? Consider the resistance F', Fig. 18.12, *Text*. (b) If the rotating shaft comes to rest from 300 rpm during 100 revolutions, what frictional work has been done? (c) What must be the diameter of the steel pin P, SAE 1020 as rolled, for a factor of safety of 6 against being sheared off? The diameter of the hub $d = 4\frac{1}{2}$ in. (d) What is the unit pressure on the face of the brake?

928. A cone clutch for industrial use is to transmit 15 hp at 400 rpm. The mean diameter of the clutch is 10 in. and the face angle $\alpha = 10°$; let $f = 0.3$ for the cast-iron cup and the asbestos lined cone; permissible $p = 35$ psi. Compute (a) the needed axial force, (b) the face width, (c) the minimum axial force to achieve engagement under load.

929. An "Airflex" clutch, Fig. 18.15, *Text*, has a 16-in. drum with a 5-in. face. This clutch is rated at 110 hp at 100 rpm with an air pressure of 75 psi. What must be the coefficient of friction if the effect of centrifugal force is neglected? (*Data courtesy of Federal Fawick Corporation.*)

930. The same as **929** except that the diameter is 6 in., the face width is 2 in., and the rated horsepower is 3.

931–940. These numbers may be used for other problems.

Section 17

WELDING

DESIGN PROBLEMS

941. A joint welded with a coated rod, is to support a steady load of 10 kips; the design is to be for a $\frac{3}{8}$-in. weld. Compute the length of weld needed for (a) a reinforced butt joint, Fig. 19.2, *Text*, (b) a lap joint, Fig. 19.3(a), (c) a T-joint, Fig. 19.3(b), (d) fillet welds, parallel loaded, Fig. 19.3(d).

942. (a) The same as **941**, except that the load is repeated and reversed for 2×10^6 cycles. (b) If design stresses from Table AT 30 are used, compute design factors based on information in Table 19.1, p. 515, *Text*. (c) Using a rational approach in accordance with the principles of Chapter 4, *Text*, determine design stresses and compare (or contrast) with those used.

943. The load F varies from 5 to 10 kips and the arrangement is such that the location is given by $m = 3$ in. and $n = 7$ in. It is desired that the weld lengths L_1 and L_2 be such that the line of action of F passes through the centroid G of the weld metal, thereby avoiding eccentric loading. Determine (a) the ratio of L_1/L_2, (b) the lengths L_1 and L_2 of $\frac{3}{8}$-in. fillet welds made with E6010 rod, for indefinite life. (c) The same as (b) except that the life expectancy is 10^5 cycles.

96

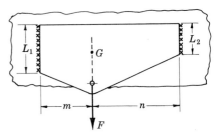

Problem 943.

944. A bracket somewhat as shown is made of structural steel and supports a

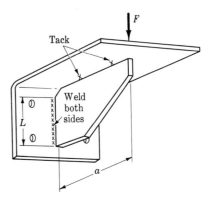

Problems 944, 958.

repeated ($R = 0$, $n_c = 2 \times 10^6$) load of 2000 lb. at a distance $a = 10$ in. from the wall. What should be the length L of a $\frac{3}{8}$-in. fillet weld that resists the entire load? Adapt the design shear stress from Table AT 30 (fillet weld).

945. A bracket is to be fabricated from flat plates by bending and welding with a shielded rod, E6010. A steady load $F = 5000$ lb., $L = 18$ in., $h = 4$ in., and $a = 6$ in. (a) Take the design shear stress for a design factor $N = 3.75$ on the ultimate shear strength, which may be estimated at 80% of s_u of rod, and find the size of fillet weld ABCDA. Compare the design stress used with values in Table AT 30. (b) Compute the thickness of the SAE 1020, rolled-steel plates if all are the same (cantilever part).

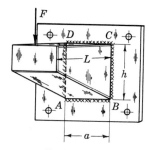

Problem 945.

946. A bracket is made with two $\frac{3}{8}$-in. steel plates A welded with a coated electrode to a vertical I-beam with fillet welds on both sides of the plate, as indicated. It supports a steady vertical load

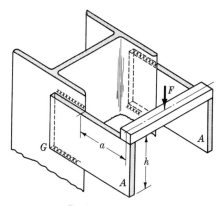

Problems 946-948.

$F = 12$ kips in a center position; $a = 14$ in., $h = 8$ in. What size and length of weld should be used? Is the stress at G the maximum one? Justify your answer.

947. The same as **946**, except that the material is aluminum alloy, welded with shielded 1100 wire, and the load is 5 kips. Let the design factor $N = 3.4$ for the information in Table AT 30; but consider other approaches, as available, to a design stress.

948. (a) Two $\frac{3}{4}$-in. plates A, arranged as shown, are to be welded with coated electrodes, E6020; $a = 12$ in., $h = 4$ in., and F repeats from 0 to 10 kips. Choose a design stress from Table AT 30 for 6×10^6 cycles and specify the size and length of weld. (b) The same as (a), except that the design is for 10^5 cycles. (c) Demonstrate that the stress at G is or is not the maximum.

949. One arm of a bracket that is to support a steady load of $F = 18$ kips without twisting is welded with an E6010 rod, as shown. The plate is 10 in. ($\approx L_2$) deep. Assume a value of L_1 (not less than 5 in.) and compute the size of fillet weld. By sketching vectors (only), compare the stress at C with that at B.

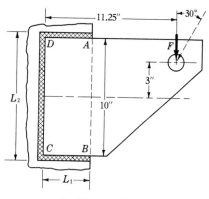

Problems 949-951.

950. The same as **949**, except that F makes a 30° angle with the vertical as indicated by the dotted line in the figure. Consider all computed stresses to be shear.

951. The same as **949**, except that the load varies from 4 to 18 kips; expected life, 2×10^6 cycles. Solve (a) by using a design stress from Table AT 30 for the given value of R, and (b) by using a design stress for $R = -1$ and the Soderberg criterion.

952. A steel plate, welded to a column as shown with E6010 rod, is to support a steady load of $F = 5$ kips, applied so as to produce no twisting of the plate; $m = 24$ in., $n = 18$ in.; the initial design is for a $\frac{3}{8}$-in. fillet weld. Compute L. Demonstrate by sketches which stress s_{sA} or s_{sD} is the larger.

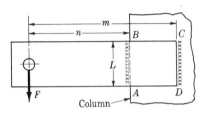

Problems 952-954, 959.

953. A steel plate, welded as shown with E6010 rod, is to support a load that varies from -5 to $+5$ kips, without twisting; $m = 14$ in., $n = 8$ in.; the initial design is for a $\frac{3}{8}$-in. fillet weld; indefinite life. Compute L.

954. The same as **953**, except that F varies from 0 to 5 kips with a life expectancy of 10^5 cycles.

955. An arm for a machine is to be fabricated by welding, coated welding rod. See figure. The left end is a hollow cylinder with $D_o = 3$ in., and it is keyed to a 2-in. shaft; $L = 14$ in., steady load $F = 600$ lb. The arm material is SAE 1020, rolled-steel plate, $\frac{1}{2}$-in. thick. Compute (a) the depth h of the arm at the hub, and (b) the size of the weld.

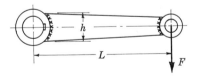

Problem 955.

956. A pair of gusset plates, $\frac{3}{8}$-in. thick, are to be welded with E6010 electrodes, as shown. The load F on the

plates varies from 0 to 10 kips (no twisting of plates). For the first approximation, assume that $BC = AD = L = 5$ in. and compute the size of weld. With freehand sketches, compare the resultant stress at each corner A, B, C, and D.

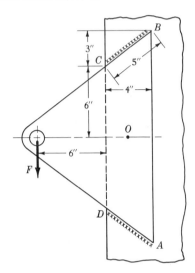

Problems 956, 957.

CHECK PROBLEMS

957. A $\frac{3}{8}$-in. gusset plate is welded with an E6010 electrode; $\frac{1}{4}$-in. fillet weld, as shown. The loading does not twist the plate and the force varies from $0.2F$ to F. For a life expectancy of 10^5 cycles, what is a safe F? Make clear how you decide upon the point of maximum stress.

958. A bracket of the type shown (**944**) is to support a load of $F = 6$ kips; $a = 8$ in., $L = 5$ in., weld size is $\frac{3}{8}$-in. (a) Determine the stress in the weld. Is this a safe value for a steady load? (b) If the welding is shielded and the load varies with $R = 0$, is the weld safe for 2×10^6 cycles? For 10^5 cycles? For Q&T alloy and 2×10^6 cycles?

959. The 1-in. plate shown (**952**) is attached with $\frac{1}{2}$-in. fillet welds, laid with E6010 rods; $L = 4$ in., $m = 15$ in., $n = 9$ in. What maximum load may be carried if it is (a) static, (b) varies from $0.5F$ to F for 2×10^6 cycles and for 10^5 cycles, (c) $R = 0$, indefinite life. (d) Considering the strengths given in Table 19.1,

Text, determine the design factor for parts (b) and (c).

960. The plate for a bracket, as shown must be welded to a member in the manner shown; $\frac{5}{16}$-in. welds with shielded arc. Compute the safe load for this plate (no twisting) (a) for static loading, (b) for $R = 0$ and indefinite life, (c) for $R = -1$ and indefinite life, (d) for $R = 0.2$ and indefinite life.

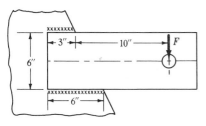

Problem 960.

961. A 2-in. round bar is welded to a vertical wall by a $\frac{3}{8}$-in. fillet weld as shown in Fig. 19.8, *Text*. The bar supports a vertical load of 800 lb. at a distance of 10 in. from the wall. What is the maximum computed stress in the weld? Would this result be safe for a varying load with $R = 0$, shielded weld?

962. The 14-in. structural-steel disk is welded to the plate by a $\frac{7}{16}$-in. fillet weld, 360°, shielded arc. The force F acts on a pin attached to the disk. The pin is short enough that the moment arm to the disk is negligible. Determine a safe force F for (a) static loading, (b) reversed loading, indefinite life, (c) a varying load from $0.3F$ to F, indefinite life.

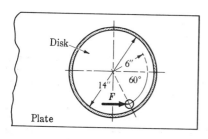

Problem 962.

963. A bracket is fabricated from $\frac{1}{2}$-in., AISI-1020 rolled plate with $\frac{3}{8}$-in. fillet welds on both sides of the plates G, H, and J as shown. The welds are made with E6016 welding rod; a central load at $L = 30$ in.; $h = 11$ in. Determine the repeated load that the welding can support.

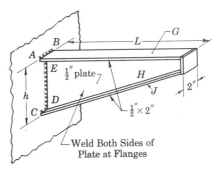

Problem 963.

964. The bracket shown is made of $\frac{1}{2}$-in., AISI-C1020 rolled plates. The $\frac{3}{8}$-in. fillet welds are on both sides of each plate A and B, E7010 welding rod. The entire bracket is normalized after welding; $L = 12$ in., $a = 8$ in., and $h = 8$ in. What is the safe maximum load if it is (a) static, (b) varies from $0.5F$ to F for 2×10^6 cycles and for 10^5 cycles, (c) $R = 0$, indefinite life. (d) Considering the strengths given in Table 19.1, *Text*, determine the design factor for parts (b) and (c).

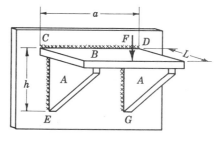

Problem 964.

965–980. These numbers may be used for other problems.

Section 18

MISCELLANEOUS PROBLEMS

THIN SHELLS, EXTERNAL PRESSURE

981. A closed cylindrical tank is used for a steam heater. The inner shell, 200 in. outside diameter and 50 ft. long, is subjected to an external pressure of 40 psi. The material is equivalent to SA 30 (ASME Pressure-Vessel Code: min. $s_u = 55$ ksi); assume an elastic limit of $s_y = s_u/2$; let $N = 5$. (a) What thickness of shell is needed from a stress standpoint? (b) For this thickness, what must be the maximum length of unsupported section to insure against collapse? (c) Choose a spacing L to give a symmetric arrangement and determine the moment of inertia of the steel stiffening rings. (d) For a similar problem, the Code recommends that $t \geq 0.76$ in., $L = 50$ in., and $I = 96$ in.[4] How do these values check with those obtained above? (e) Without stiffening rings, what thickness would be needed?

982. The same as **981**, except that $p = 175$ psi, $D = 4$ ft., and the length of the tank is 18 ft.

983. A closed cylindrical tank, 6 ft. in diameter, 10 ft. long, is subjected to an internal pressure of 1 psi absolute. The atmospheric pressure on the outside is 14.7 psi. The material is equivalent to SA 30 (ASME Pressure-Vessel Code: min.

$s_u = 55$ ksi); assume an elastic limit of $s_y = s_u/2$; let $p_c/p = 5$. (a) What thickness of shell is needed for the specified design stress? (b) For this thickness, what must be the maximum length of unsupported section to insure against collapse? (c) Choose a symmetric spacing L of stiffening rings, and compute their moment of inertia and the cross-sectional dimensions h and b if they are rectangular with $h = 2b$. (d) Suppose that the tank had no stiffening rings. What thickness of shell would be needed? What is the approximate ratio of the weight of this shell to the weight of the shell found in (a)? Material costs are roughly proportional to the weight.

STEEL TUBES, EXTERNAL PRESSURE

984. A long lap-welded steel tube, 8-in. OD, is to withstand an external pressure of 120 psi. with $N = 5$. (a) What should be the thickness of the wall of the tube? (b) What is the ratio D/t? Is it within the range of the Stewart equation? (c) Assuming the internal pressure to be negligible relative to the external pressure, calculate the maximum principal stress from equation (8.13), p. 255, *Text*. What design factor is given

100

by this stress compared with s_y for AISI C1015, annealed? (d) Compute the stress from the thin shell formula.

985. A long lap-welded steel tube, 3-in. OD, is to withstand an external pressure of 150 psi with $N = 5$. Parts (a)–(c) are the same as in **984.**

986. A long lap-welded tube, 3-in. OD, is made of SAE 1015, annealed. Let the shell thickness $t = D/40$ and $N = 5$. (a) What is the corresponding safe external pressure? (b) Compute the maximum principal stress (p. 255, *Text*), assuming a negligible internal pressure. What design factor is given by this stress compared with s_y? (c) Compare with the stress computed from the thin-shell formula.

FLAT PLATES

987. A circular plate 24 in. in diameter and supported but not fixed at the edges, is subjected to a uniformly distributed load of 125 psi. The material is SAE 1020, as rolled, and $N = 2.5$ based on the yield strength. Determine the thickness of the plate.

988. The cylinder head of a compressor is a circular cast-iron plate (ASTM class 20), mounted on a 12-in. cylinder in which the pressure is 250 psi. Assuming the head to be supported but not fixed at the edges, compute its thickness for $N = 6$ based on ultimate strength.

989. A 10×15-in. rectangular opening in the head of a pressure vessel, whose internal pressure is 175 psi, is covered with a flat plate of SAE 1015, annealed. Assuming the plate to be supported at the edges, compute its thickness for $N = 6$ based on ultimate strength.

CAMS

990. The force between a $\frac{5}{8}$-in. hardened steel roller and a cast-iron (140 BHN) cam is 100 lb.; radius of cam curvature at this point is $1\frac{1}{4}$ in. Compute the contact width.

991. A radial cam is to lift a roller follower 3 in. with harmonic motion during a 150° turn of the cam; $1\frac{1}{2}$-in. roller

of hardened steel. The reciprocating parts weigh 10 lb., the spring force is 175 lb., the external force during the lift is 250 lb. The cast-iron (225 BHN) cam turns 175 rpm. The cam curvature at the point of maximum acceleration is $1\frac{1}{2}$-in. radius. Compute the contact width.

992. The same as **991**, except that the motion of the follower is cycloidal.

993. The same as **991**, except that the motion of the follower is parabolic.

FLYWHEELS AND DISK

994. A cast-iron flywheel with a mean diameter of 36 in. changes speed from 400 rpm to 380 rpm while it gives up 8000 ft-lb. of energy. What is the coefficient of fluctuation, the weight, and the approximate sectional area of the rim?

995. The energy required to shear a 1-in. round bar is approximately 1000 ft-lb. In use, the shearing machine is expected to make a maximum of 40 cutting strokes a minute. The frictional losses should not exceed 15% of the motor output. The shaft carrying the flywheel is to average 150 rpm. (a) What motor horsepower is required? (b) Assuming a size of flywheel and choosing an appropriate C_f, find the mass and sectional dimensions of the rim of a cast-iron flywheel. The width of the rim is to equal the depth and is not to exceed $3\frac{1}{2}$ in. It would be safe to assume that all the work of shearing is supplied by the kinetic energy given up by the flywheel.

996. The same as **995**, except that the capacity of the machine is such as to cut $1\frac{1}{2}$-in. round brass rod, for which the energy required is about 400 ft-lb./ sq. in. of section.

997. A 75-hp Diesel engine, running at 517 rpm, has a maximum variation of output of energy of 3730 ft-lb. The engine has three $8 \times 10\frac{1}{2}$-in. cylinders and is directly connected to an a-c generator. (a) What should be the weight and sectional area of the flywheel rim if it has an outside diameter of 48 in.? (b) The actual flywheel and generator have $Wk^2 = 6787$ lb-ft.[2] Compute the corresponding coefficient of fluctuation and compare.

998. A 4-ft. flywheel, with a rim 4 in. thick and 3 in. wide, rotates at 400 rpm. If there are 6 arms, what is the approximate stress in the rim? Is this a safe stress? At what maximum speed should this flywheel rotate if it is made of cast iron, class 30?

999. A hollow steel shaft with $D_o = 6$ in. and $D_i = 3$ in. rotates at 10,000 rpm. (a) What is the maximum stress in the shaft due to rotation? Will this stress materially affect the strength of the shaft? (b) The same as (a), except that the shaft is solid.

1000. A circular steel disk has an outside diameter $D_o = 10$ in., and an inside diameter $D_i = 2$ in. Compute the maximum stress for a speed of (a) 10,000 rpm, (b) 20,000 rpm. (c) What will be the maximum speed without danger of permanent deformation if the material is AISI 3150, OQT at 1000°F?

1001. The same as **1000**, except that $D_i = 1$ in.

1002. A circular steel disk, with $D_o = 8$ in. and $D_i = 2$ in., is shrunk onto a solid steel shaft with an interference of metal $i = 0.002$ in. (a) At what speed will the pressure in the fit become zero as a result of the rotation? Assume that the *shaft* is unaffected by centrifugal action. (This effect is relatively small.) (b) Compute the maximum stress in the disk and the pressure at the interface when the speed is 10,000 rpm. *Note:* The maximum stress in the disk σ_{th} is obtained by adding equations (8.15) of § 8.26, *Text*, and (**n**) of § 20.9. The resulting equation together with equation (**s**) of § 8.27 can then be used to obtain p_i and σ_{th}; where $\sigma_{ts} = -p_i$ for a solid shaft.

NOTE. *There are not 1002 problems in this book; see blank numbers at the end of each section for the use of the instructor.*

APPENDIX

The material of this appendix is the same in content and arrangement as in the fourth edition of *Design of Machine Elements*, by V. M. Faires, except that in addition, there is a chart for aiding in making trial solutions of Buckingham's dynamic-load equation (p. 156), a table of decimal equivalents of an inch at the end, preferred sizes (p. 107), "standard" gear-tooth pitches (p. 143), a few useful mathematical relations (p. 154), and a number of selected equations taken from the *Text*. The sequence of the tables and charts is closely the sequence of subject matter in the *Text*.

The student should familiarize himself with the following contents, because he may be allowed the use of this material, and not the text, on examinations and because this material may often be more convenient.

TABLE AT 1 PROPERTIES OF SECTIONS

I_x = moment of inertia about the axis x–x, J=polar moment of inertia about the centroidal axis, $Z = I/c$ = rectangular section modulus, about x–x, $Z' = J/c$ = polar section modulus, $k = \sqrt{I/\text{area}}$ = radius of gyration.

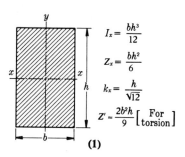

$$I_x = \frac{bh^3}{12}$$

$$Z_x = \frac{bh^2}{6}$$

$$k_x = \frac{h}{\sqrt{12}}$$

$$Z' = \frac{2b^2h}{9}\left[\begin{array}{c}\text{For}\\\text{torsion}\end{array}\right]$$

(1)

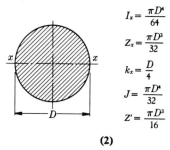

$$I_x = \frac{\pi D^4}{64}$$

$$Z_x = \frac{\pi D^3}{32}$$

$$k_x = \frac{D}{4}$$

$$J = \frac{\pi D^4}{32}$$

$$Z' = \frac{\pi D^3}{16}$$

(2)

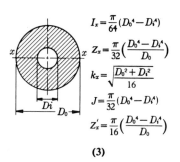

$$I_x = \frac{\pi}{64}(D_o^4 - D_i^4)$$

$$Z_x = \frac{\pi}{32}\left(\frac{D_o^4 - D_i^4}{D_o}\right)$$

$$k_x = \sqrt{\frac{D_o^2 + D_i^2}{16}}$$

$$J = \frac{\pi}{32}(D_o^4 - D_i^4)$$

$$Z_x' = \frac{\pi}{16}\left(\frac{D_o^4 - D_i^4}{D_o}\right)$$

(3)

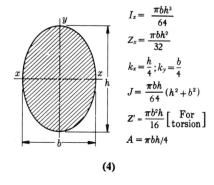

$$I_x = \frac{\pi bh^3}{64}$$

$$Z_x = \frac{\pi bh^2}{32}$$

$$k_x = \frac{h}{4}; k_y = \frac{b}{4}$$

$$J = \frac{\pi bh}{64}(h^2 + b^2)$$

$$Z' = \frac{\pi b^2 h}{16}\left[\begin{array}{c}\text{For}\\\text{torsion}\end{array}\right]$$

$$A = \pi bh/4$$

(4)

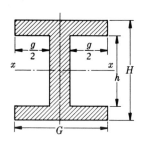

$$I_x = \frac{1}{12}(GH^3 - gh^3)$$

$$Z_x = \frac{GH^3 - gh^3}{6H}$$

$$k_x = \sqrt{\frac{1}{12}\left[\frac{GH^3 - gh^3}{GH - gh}\right]}$$

(5)

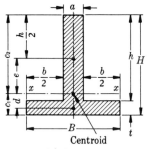

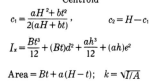

Centroid

$$c_1 = \frac{aH^2 + bt^2}{2(aH + bt)}, \qquad c_2 = H - c_1$$

$$I_x = \frac{Bt^3}{12} + (Bt)d^2 + \frac{ah^3}{12} + (ah)e^2$$

$$\text{Area} = Bt + a(H - t); \quad k = \sqrt{I/A}$$

(6)

105

TABLE AT 2 MOMENTS AND DEFLECTIONS IN BEAMS

F lb. = applied force; w = pounds per inch of length; $F = wL$, where L in. = length; E psi = modulus of elasticity, tension; I in.^{4} = moment of inertia; y in. = deflection; θ radians = slope. For other beams of uniform strength, see § 6.24.

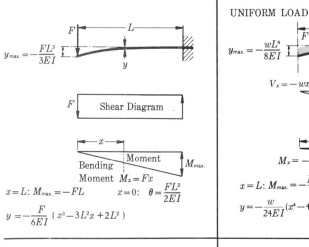

$$y_{max.} = -\frac{FL^3}{3EI}$$

F | Shear Diagram

Bending Moment $M_x = Fx$

$x = L: M_{max.} = -FL \qquad x = 0: \theta = \frac{FL^2}{2EI}$

$$y = -\frac{F}{6EI}(x^3 - 3L^2x + 2L^2)$$

UNIFORM LOAD

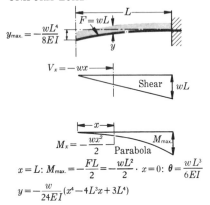

$$y_{max.} = -\frac{wL^4}{8EI}$$

$V_x = -wx$ — Shear — wL

$$M_x = -\frac{wx^2}{2} \quad \text{Parabola} \quad M_{max.}$$

$x = L: M_{max.} = -\frac{FL}{2} = -\frac{wL^2}{2} \cdot x = 0: \theta = \frac{wL^3}{6EI}$

$$y = -\frac{w}{24EI}(x^4 - 4L^3x + 3L^4)$$

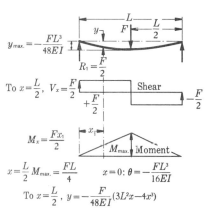

$$y_{max.} = -\frac{FL^3}{48EI}$$

$R_1 = \frac{F}{2}$

To $x = \frac{L}{2}, \quad V_x = \frac{F}{2}$ Shear $-\frac{F}{2}$

$$M_x = \frac{Fx_1}{2}$$

$M_{max.}$ Moment

$x = \frac{L}{2} \; M_{max.} = \frac{FL}{4} \qquad x = 0: \theta = -\frac{FL^3}{16EI}$

To $x = \frac{L}{2}, \quad y = -\frac{F}{48EI}(3L^2x - 4x^3)$

UNIFORM LOAD

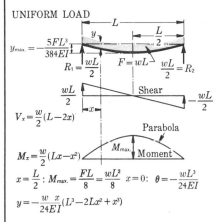

$$y_{max.} = -\frac{5FL^3}{384EI}$$

$R_1 = \frac{wL}{2} \qquad F = wL \qquad \frac{wL}{2} = R_2$

$\frac{wL}{2}$ Shear $-\frac{wL}{2}$

$$V_x = \frac{w}{2}(L - 2x)$$

Parabola

$$M_x = \frac{w}{2}(Lx - x^2)$$

$M_{max.}$ Moment

$x = \frac{L}{2}: M_{max.} = \frac{FL}{8} = \frac{wL^2}{8} \quad x = 0: \theta = -\frac{wL^3}{24EI}$

$$y = -\frac{w\,x}{24EI}(L^3 - 2Lx^2 + x^3)$$

FIXED ENDS

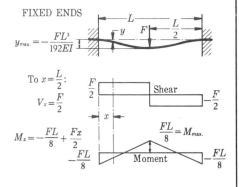

$$y_{max.} = -\frac{FL^3}{192EI}$$

To $x = \frac{L}{2}:$

$V_x = \frac{F}{2}$

$\frac{F}{2}$ Shear $-\frac{F}{2}$

$$M_x = -\frac{FL}{8} + \frac{Fx}{2}$$

$\frac{FL}{8} = M_{max.}$

$-\frac{FL}{8}$ Moment $-\frac{FL}{8}$

UNIFORM LOAD

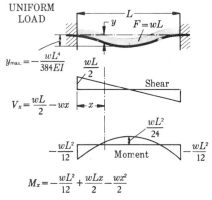

$$y_{max.} = -\frac{wL^4}{384EI}$$

$\frac{wL}{2}$

$\frac{wL}{2}$ Shear

$V_x = \frac{wL}{2} - wx$

$\frac{wL^2}{24}$

$-\frac{wL^2}{12}$ Moment $-\frac{wL^2}{12}$

$$M_x = -\frac{wL^2}{12} + \frac{wLx}{2} - \frac{wx^2}{2}$$

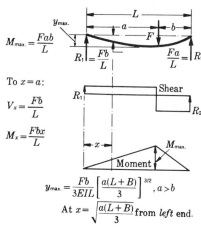

$M_{max.} = \dfrac{Fab}{L}$

$R_1 = \dfrac{Fb}{L}$ $\dfrac{Fa}{L} = R_2$

To $x = a$:

$V_x = \dfrac{Fb}{L}$

$M_x = \dfrac{Fbx}{L}$

$y_{max.} = \dfrac{Fb}{3EIL} \left[\dfrac{a(L+B)}{3} \right]^{3/2}, a > b$

At $x = \sqrt{\dfrac{a(L+B)}{3}}$ from *left* end.

$0 < x < a:$ $y = \dfrac{-Fbx}{6EIL}(L^2 - b^2 - x^2)$

$a < x < L:$ $y = -\dfrac{Fa(L-x)}{6EIL}\left[L^2 - a^2 - (L-x)^2 \right]$

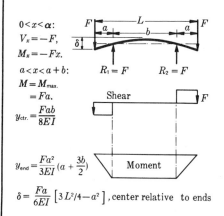

$0 < x < \alpha:$
$V_x = -F,$
$M_x = -Fx.$

$a < x < a + b:$ $R_1 = F$ $R_2 = F$

$M = M_{max.}$
$= Fa.$

$y_{ctr.} = \dfrac{Fab}{8EI}$

$y_{end} = \dfrac{Fa^2}{3EI}\left(a + \dfrac{3b}{2}\right)$

$\delta = \dfrac{Fa}{6EI}\left[3L^2/4 - a^2 \right],$ center relative to ends

UNIFORM STRENGTH, CANTILEVER

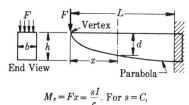

$M_x = Fx = \dfrac{sI}{c}$. For $s = C$,

$d^2 = \dfrac{6F}{bs}x = \dfrac{x}{L}h^2$

At $x = 0$: $y_{max.} = -\dfrac{8FL^3}{bEh^3}$

UNIFORM STRENGTH, SIMPLE BEAM

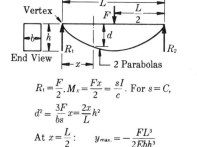

$R_1 = \dfrac{F}{2} \cdot M_x = \dfrac{Fx}{2} = \dfrac{sI}{c}$. For $s = C$,

$d^2 = \dfrac{3F}{bs}x = \dfrac{2x}{L}h^2$

At $x = \dfrac{L}{2}$: $y_{max.} = -\dfrac{FL^3}{2Ebh^3}$

PREFERRED SIZES

$\frac{1}{64}$ between $\frac{1}{64}$–$\frac{1}{32}$; $\frac{1}{32}$ between $\frac{1}{32}$–$\frac{3}{16}$;
$\frac{1}{16}$ between $\frac{3}{16}$–$\frac{7}{8}$; $\frac{1}{8}$ between $\frac{7}{8}$–3;
$\frac{1}{4}$ between 3–6; $\frac{1}{2}$ above 6.

TABLE AT3 TYPICAL PROPERTIES OF OF SOME NONFERROUS METALS[0.1,2.1,2.8,2.9,2.12]

For *Aluminum alloys*, let Poisson's ratio $\mu = 0.33$; torsional ult. $= 0.65s_u$; torsional yield str. $= 0.55s_y$. Extruded forms over $\frac{3}{4}$ in. will have s_u about 15% greater.

For *Magnesium alloys*, let the flexural strength (symmetric sections) be the average of the tensile and compressive strengths. See note (k). Let $\mu = 0.35$.

Abbreviations: H, hard; $\frac{1}{4}$H, $\frac{1}{4}$ hard; H14, temper designation meaning $\frac{1}{2}$ hard; HT, heat treated; T4, temper designation meaning solution heat treated.

Notes: (a) At 0.5% total elongation under load. (b) Cold reduction of 11%. (c) At 0.2% offset. (d) BHN. (e) BHN with 500 kg. load. (f) Minimum. (g) Flat 0.04 in. thick. (h) End.

MATERIAL (ASTM No.)	CONDITION	ULT. STR., ksi s_u	s_{us}	TEN YD. s_y ksi	END. STR., s_n ksi at No. cycles
COPPER ALLOYS					
Admiralty brass (B111)	Annealed	53		22(a)	18 at 10^7
Aluminum bronze (B150-2)	Extruded rod (b)	120		70(a)	32 at 3×10^8
Aluminum bronze (B150-1)	Annealed, 800°F	100			28 at 8×10^7
Beryllium copper (B194)	HT	175		130	35 at 10^8(h)
Cartridge brass (B134-6)	$\frac{1}{2}$H, rod	70	42	52(a)	22 at 5×10^7
Commercial bronze (B134-2)	Spring H(g)	72	42	62(a)	21 at 15×10^6
Free-cutting brass (B16)	2″ Rod, $\frac{1}{2}$H	55	32	44(a)	20 at 10^8
Manganese bronze (B138-A)	$\frac{1}{2}$H, rod	75	48	65(a)	30 at 10^8(g)(q)
Naval brass (B124-3)	$\frac{1}{4}$H, rod	70	43	48(a)	15 at 3×10^8(b)
Nickel silver B	H(g)	100		85(a)	19
Phosphor bronze (B139C)	$\frac{1}{2}$H, rod	80		65(a)	31 at 10^8(r)
Silicon bronze (B98-B)	H, 1″ rod	65	45	35(a)	25 at 10^8
Yellow brass (B36-8)	$\frac{1}{8}$H, 1″ rod	55	36	40(a)	11 at 10^8(z)
Yellow brass (B36-8)	$\frac{1}{2}$H (g)	61	40	50(a)	
ALUMINUM ALLOYS					
3003-H14 (M1A)	Strain hard.	22	14	21(c)	9 at 5×10^8
2014-T6 (CS41A)	HT, aged	70	42	60(c)	18 at 5×10^8
2024-T4 (CG42A)	HT, aged	68	41	47(c)(p)	20 at 5×10^8
6061-T6 (p) (GS11A)	HT, aged	45	30	40(c)	14 at 5×10^8
7075-T6 (ZG62A)	HT, aged	82	48	72(c)	23 at 5×10^8
360	Die casting	40	27	24(c)	17 at 5×10^8
355 T6	Sand casting	35	28	25(c)	9 at 5×10^6
MAGNESIUM ALLOYS					
AZ61A-F	Extruded bar	45	20	33(c,k)	17 at 5×10^8
AZ80A-T5	Forged, aged	50	23	36(c,k)	16 at 5×10^8
AZ91C-T6	Sand casting	40		19(c,k)	14 at 5×10^8 (z)
LEAD AND TIN ALLOYS					
Babbitt (B23-46T-8)	At 68°F (l)	10	$s_{yc} = 3.4$ (m)		3.9 at 2×10^7
Babbitt (B23-46T-8)	At 212°F (l)	5.4	$s_{yc} = 1.7$ (m)		
Tin babbitt (B23-49-1)	At 68°F (l)	9.3	$s_{yc} = 4.4$ (m)		3.8 at 2×10^7
MISCELLANEOUS					
Hastelloy B	Sand cast	90		50(c)	66 at 10^8(n)
K Monel	Cold dr., aged	140	98	100(c)	42 at 10^8
Platinum alloy	Annealed	45			
Titanium (B265, gr. 5) (t)	Annealed (s)	135		130(c)	42 (f)
Titanium (B265, gr. 5) (t)	Hardened (s)(v)	170		158(c)	61 (f))
Zinc (AC41A)	Die cast (w)	47.6	38		8 at 10^8

strength in reversed torsion, 25 ksi. (i) α in./in.–°F, coef. of thermal expansion; room temp. (j) Varies with size of test specimen. (k) Yield point in compression; alloy AZ91C–T6, 19 ksi; AZ61A–F, 19 ksi; AZ80A–T5, 28 ksi. (l) Chill cast. (m) In compression, at 0.125% set. (n) At 1200°F, after water quench and aging. (o) Estimated. (p) Used for rolled structural shapes. (q) For manganese gear bronze, use $s_n = 17$ ksi. (r) For phosphor gear bronze, SAE 65, use $s_n = 24$ ksi. (s) Normal temp.; see § 2.21. (t) Sheet. (u) About 17.5 in compression. (v) Water quenched and aged at 975°F. (w) $s_{uc} = 87$ ksi, Charpy = 48. (x) After 1 yr. (y) Pure platinum. (z) Die castings.

MOD. EL. $E \times 10^{-6}$	MOD. EL. SH. $G \times 10^{-6}$	ELONG. % 2 in. (j)	ROCK. HARD.	DENSITY $lb/in.^3$	$\alpha \times 10^6$ (i)	PERCENTAGES OF ELEMENTS
15	5.8	65	F75	0.308	11.2	71 Cu, 28 Zn, 1 Sn
16	6.5	12	B100	0.274	9	81.5 Cu, 9.5 Al, 5 Ni, 2.5 Fe, 1 Mn
15	6.5	25	B90	0.274	9.2	91 Cu, 9 Al
19	7.3	5	C37	0.297	9.3	1.9 Be, 0.2 Ni or Co
16	6	30	B80	0.308	11.1	70 Cu, 30 Zn
17	6.4	3	B78	0.318	10.2	90 Cu, 10 Zn
14	5.3	32	B75	0.307	11.4	61.5 Cu, 35.5 Zn, 3 Pb
16	6	25	B80	0.302	11.8	58 Cu, 39 Zn, plus
15	5.6	25	B80	0.304	11.8	60 Cu, 39.25 Zn, 0.75 Sn
18		3	B91	0.314	9.3	55 Cu, 27 Zn, 18 Ni
16	6	33	B85	0.318	10.1	92 Cu, 8 Sn
17	6.4	10	B80	0.316	9.9	97.7 Cu, 1.5 Si, plus
15	5.6	48	B55	0.306	11.3	65 Cu, 35 Zn
15	5.6	23	B70	0.306	11.3	65 Cu, 35 Zn
10	3.85	16	40(e)	0.099	12.9	1.0 Mn, others
10.6	4.0	13	135(e)	0.101	12.8	3.9 Cu, 0.5 Si, 0.4 Mn, 0.2 Mg
10.6	4.0	20	120(e)	0.098	13.0	3.8 Cu, 1.2 Mg, 0.3 Mn
10	3.75	17	95(e)	0.100	12.7	0.15 Cu, 0.8 Mg, 0.4 Si
10.4	3.9	10	150(e)	0.101	12.9	5.1 Zn, 2.1 Mg, 1.2 Cu
10.3	3.85	1.8	70(e)	0.095	11.7	9 Si, 0.4 Mg
10.3	3.85	3	80(e)	0.098	11.7	1 Cu, 4.5 Si, 0.4 Mg
6.5	2.4	16	E72	0.065	14.4	6 Al, 1 Zn, 0.2 Mn
6.5	2.4	6	E82	0.065	14.4	8.5 Al, 0.5 Zn, 0.15 Mn
6.5	2.4	5	E77	0.066	14.4	9 Al, 0.7 Zn, 0.2 Mn
4.2		5	20(e)	0.36	13.3	80 Pb, 15 Sb, 5 Sn
		27	10(e)	0.36	13.3	80 Pb, 15 Sb, 5 Sn
7.3		2	17(e)	0.265		91 Sn, 4.5 Sb, 4.5 Cu
26.5		10	B93	0.334	5.55	62 Ni, 28 Mo, 5 Fe
26		20	C30	0.306	7.8	66 Ni, 29 Cu, 3 Al
		35	90(e)	0.722	5.0(y)	10 Rhodium
15(u)		12		0.160	5.8	⎰ 6 Al, 4 V
15(u)		7		0.160	5.8	⎱ Hi. temp. aero. service
2(x)		7	91(e)	0.24	15.2	4 Al, 1 Cu, 0.04 Mg

TABLE AT 4 TYPICAL PROPERTIES
OF SOME STAINLESS STEELS[2.1,2.3,2.11,2.18]

Notes: (a) Coef. of thermal expansion near room temp., α in./in.-°F. (b) Approx. average values of ult. strength of 403, 410, and 416 are given by $s_u = 5 + 0.465$ (BHN) in ksi.[2.3] (c) Varies with details of heat treatment and cold working. (d) Cold worked, full hard. (e) Endurance limits for stainless steels may be estimated at $0.4s_u$, up to tensile strength

MATERIAL AISI No.	ULT. STR. s_u, ksi (c)	TEN. YD. s_y, ksi (c)	END. LIM. s_n' (e)	MOD. EL. $E \times 10^{-6}$ (f)	ELONG. 2 in.% (c)	RED AREA,% (c)
301, $\frac{1}{4}$ hard . . .	125(h)	75(h)	30(g)	28	25(h)	
302, annealed . . .	90	37	34	28	57	65
302, $\frac{1}{4}$ hard . . .	125(g)(h)	75(h)	70(d)	28	12(h)	
303, annealed . . .	90	35	35	28	50	55
304, annealed . . .	85	35		28	50	70
316, cold worked (i) . .	90	60	40	28	45	65
321, annealed . . .	87	35	38	28	50	65
347, annealed . . .	90	40	39	28	50	65
403, 410, heat treated (b)	110(h)	85(h)	58	29	20	65
410, cold worked (b) .	100(h)	85	53	29	17	60
416, annealed (b) . .	75	40	40	29	30	60
430, annealed . . .	75	45	40	29	25	65
431, OQT 1000 (b) . .	150	130		29	18	60
17-7 PH rod (j) . .	175	155	41	29	6(h)	34

TABLE AT 5 TYPICAL PROPERTIES
OF A FEW PLASTICS[2.19,2.21,2.27]

Notes: (a) TS, thermosetting; TP, thermoplastic. (b) National Electrical Mfg. Assoc. grades. (c) Flatwise. (d) For $\frac{1}{3}$ to 1-in. dia. Reduce 15% for sizes 1 to 2 in. (e) For $\frac{1}{8}$ to 1-in. dia. Reduce 10% for sizes 1 to 2 in. (f) Min. values. (g) Bending strength, symmetric sections. (h) Specific gravity. (i) Aver. water absorption, 24-hr., $\frac{1}{4}$-in. thickness, per cent.

MATERIAL	TYPE (a)	CONDITION (k)	ULT. STR., s_u ksi	COMP. ULT. STR., s_{uc}	FLEX STR., s_f (g) ksi
Phenol-formaldehyde					
Grade X (b) (l) . .	TS	L. sheet	14	35	23
Grade XX (b) (l) . .	TS	L. rod	8.5(d)	20	15(e)
Grade C (b) (l) . . .	TS	L. rod	7.5(d)	20	17(e)
Grade A (b) (l) . .	TS	L. rod	6(d)	15	10(e)
Wood flour filler (p) .	TS	M	6(f)	24	9(f)
Urea-formaldehyde . .	TS	M	9	25	10
Polyvinyl chloride . .	TP	M	8	10	
Polyvinyl chloride (n) . .	TP	M	8	13	
Polymethyl Methacrylate .	TP	M	8	14	9
Polystyrene (f) . . .	TP	M	5	11.5	6
Polyamide (m) . . .	TP	M	11.8(s)	4.9(q)	13.8
Cellulose Acetate . . .	TP	M	4.5	20	
Polyethelene (f) . . .	TP	M	1.7	0.4(m)	1.7
Polytetrafluoroethylene (m)	TP	M	3.8(u)	1.8(t)	2
Polyvinylidene chloride .	TP	M	5	2.4	
Polychlorotrifluoroethylene	TP	M	6	5	

of about 160 ksi[2.18] (f) Varies some with condition: annealed, cold worked, stress relieved. In shear, for cold drawn spring wires, $G \approx 10.6 \times 10^6$ psi. (g) 0.058-in. strip. (h) Minimum. (i) 1-in. bars. (j) PH, precipitation hardened; Republic Steel TH 1050; guaranteed $s_{umin} = 170$ ksi; s_n' at 10^8.

BHN (aver.) (c)	DENSITY $lb/in.^3$	IZOD ft-lb. (c)	$\alpha \times 10^6$, (a)	REMARKS
260	0.286		9.4	(17% Cr, 7% Ni) General use; trim, structural.
150	0.286	90	9.6	Austenitic. Hardenable by cold work.
260	0.286		9.6	302, 303 are 18-8 stainless steels.
160	0.286	80	9.6	Austenitic. Hardenable by cold work.
150	0.286	110	9.6	Austenitic. Hardenable by cold work.
190	0.286		8.9	Austenitic. Hardenable by cold work.
150	0.290	110	9.3	Stabilized by Ti.
160	0.286	100	9.3	Austenitic. Hardenable by cold work.
225	0.279	75	5.7	Martensitic. Hardenable by HT.
205	0.279	80	5.7	Martensitic. Max. hardness.
155	0.278	70	5.7	Martensitic. Hardenable by HT.
160	0.277	35	5.8	Ferritic. Not hardened by HT.
325	0.28	50	6.5	Martensitic. Hardened by HT to high strength.
390	0.276		5.6	(17% Cr, 7% Ni, 1.15% Al) Solution annealed, etc.

(j) 48-hr. immersion. (k) L, laminated; M, molded. (l) When used for gears, let $s_n = 6000$ psi. (m) Yield strength. (n) Unplasticized. (p) General purpose. (q) At 1% def. (r) At failure. (s) At 73°F. (t) At 5% strain. (u) Rupture.

ELONG. % (r)	ROCK. HARD.	MOD. EL., $E \times 10^{-5}$	SP. GR. (h)	IZOD, ft-lb. (f)	%H$_2$O Ab-sorp.(i)	A FEW TRADE NAMES®
	M100	4–20	1.35	1.3(c)	1.4	Bakelite, Durez, Formica,
	M100	4–20	1.35	1.0(c)	0.65	Textolite, Micarta, Synthane,
	M100	3.5–15	1.35	3.2(c)	1.2	Durite.
	M90	3.5–15	1.65	1.8(c)	0.65	
0.4–0.8	M100	10	1.4	0.4	0.8	
0.6	M118	15	1.45	0.24	0.4	Beetle, Sylplast, Plaskon.
30	R65	3	1.2	0.8	0.05	Geon, Vinylite, Marvinol.
10	M70	8	1.41	0.4	0.1	Exon, Pliovic, Ultron.
8	M100	4	1.16	0.4	0.3(j)	Lucite, Plexiglass, Perspex.
1.2	M85	0.5	1.06	0.2	0.03	Lustrex, Styron, Styrene, Pliolite.
60(s)	R118	3.5	1.14	0.9(s)	1.5	Nylon, Zytel, 101.
20	R100	2	1.27	4	1.5–2.9	Plastacele, Celanese, Kodapak.
30–500	R11	0.15	0.92		0.01	Dylan, Alathon, Orizon.
100–200	R20	0.6	2.2	2–4	none	Teflon (TFE).
200	M55	0.7	1.7	0.7	0.1	Saran.
200	R110	2.5	2.1	4	none	Kel-F, Fluorothene.

TABLE AT 6   TYPICAL PROPERTIES OF CAST FERROUS METALS[2.1,2.8,2.14,2.16,2.17]

Notes: Approximate *coefficients of thermal expansion*, in./in.-°F are: gray iron, 5.6 × 10^{-6}; malleable iron, 6.6 × 10^{-6}; nodular iron, 6.7 × 10^{-6}; cast steel, 6.5 × 10^{-6} (but varies significantly with composition). *Poisson's ratio*: gray iron, 0.211 (min.); malleable iron, 0.265; nodular iron, 0.16; cast steel, 0.27. (a) ASTM and SAE specifications are not identical. (b) Machinability, relative values, AISI B1112 = 100%. (c) 1.2-in. dia., 18-in. supports. (d) Test results suggest that the flexural strength of cast iron in *symmetric* sections, computed from $s_f = M/Z$, is about $1.9s_u$ to $2s_u$. Use $1.9s_u$. (e) Estimated. (f) Minimum values. Typical values may range 10–40% higher. (g) ASTM 35 and higher are considered to be high-strength, and are definitely more expensive. (h) For cast iron, at 25% of ult. stress; varies with section size and chemical analysis. (i) Reversed bending. For

MATERIAL, SPEC. NO.	ULT. STRENGTH, *ksi*				TRANSV. STRENGTH lb. (c)	END. LIM. s'_n, *ksi* (i)	TEN. YD., s_y *ksi*
	s_u	s_{uc}	s_{us}	*Tors.*			
GRAY IRON (g) (as cast)	(d)	(d)				(e)	
ASTM SAE(a) .							
20 110 . .	20(f)	83	32	26	1850	10	
25	25(f)	97	35	32	2175	11.5	
30 111 . .	30(f)	109	41	40	2525	14	
35(g) 120 . .	35(f)	124	49	48.5	2850	16	
40(g) 121 . .	40(f)	140	52	57	3175	18.5	
50(g) . . .	50(f)	164	64	73	3600	21.5	
60(g) . . .	60(f)	187	60	88.5	3700	24.5	
Ni-Resist, Inco K-6 .	25(f)	100(f)					
Meehanite, (w) . .	35(f)						28(f)
MALLEABLE IRON							
ASTM Grade							
A47-52 32 510 .	52	(o)	48	58		25.5	34
A47-52 35 018 .	55	(o)	43	58		27	36.5
NODULAR CAST IRON (j)							
60-45-10 (annealed) (q) .	70	(s)		57		30	55
80-60-03 (as cast) (p) .	88	(s)		73		40	65
100-70-03 (heat treat) (r)	110	(s)		88(e)		44(e)	80
CAST STEEL					MAX. CARBON AND HEAT TREAT.		
ASTM SAE(a)							
A27-58(t). . . .	60(f)	60(f)	0.3% C, Annealed			25	30(f)
A27-58(t) 0030(k) .	65(f)	65(f)	0.3% C, Normalized			28	35(f)
A27-58	70(f)	70(f)	0.35% C, Normalized			31	36(f)
A27-58	70(f)	70(f)	0.25% C, Normalized				40(f)
A148-58 080 . .	80(f)	80(f)	N&T			35	40(f)
A148-58 . . .	80(f)	80(f)	WQT			35	50(f)(v)
A148-58 090 . .	90(f)(v)	90(f)	N&T			41	60(f)
A148-58 0105 . .	105(f)	105(f)	WQT			49	85(f)
A148-58	120(f)	120(f)	WQT			55	95(f)(w)
A148-58 0150 . .	150(f)	150(f)	WQT			65	125(f)
A148-58 0175 . .	175(f)	175(f)	WQT			77	145(f)

gray iron, $0.4s_u < S_n' < 0.6s_u$. (j) The number indicates minimum properties; e.g., 80-60-03 indicates $s_u = 80$ ksi, $s_y = 60$ ksi (0.2% offset), and 3% elongation, minimum, in approximately 1-in. section. (k) 0.3% C, max. (l) N&T, normalized and tempered. Properties of steel castings vary with carbon and alloy contents and with heat treatment, as for wrought steel; min. $s_n' \approx 0.4s_u$. (m) Charpy impact, keyhole notch, 70°F, ft-lb. (n) Charpy impact, V-notched. (o) Take as equal to s_u. (p) ASTM A339-55. (q) ASTM A395-56T. (r) ASTM A396-58. (s) For design, assume compressive ultimate and yield strengths of nodular iron to be the same as s_u and s_y, respectively. (t) Common commercial grades. (u) Tempered at 1200°F. (v) Typical $s_u \approx 96$ ksi, $s_y = 73$ ksi when WQT 1200. (w) General purpose type.

MOD. ELAS., psi $E \times 10^{-6}$	SHEAR MOD., psi $G \times 10^{-6}$	BHN	IZOD ft-lb.	DENSITY lb/in.³	MCH. (b)	REC. MIN. WALL THICK.	
(h)							
9.6(f)	3.9(f)	156		0.253		$t = \frac{1}{8}$ in.	
11.5(f)	4.6(f)	174		0.253		$t = \frac{1}{4}$ in.	
13(f)	5.2(f)	201	23	0.254	80	$t = \frac{3}{8}$ in.	
14.5(f)	5.8(f)	212	25	0.257	65	$t = \frac{3}{8}$ in.	
16(f)	6.4(f)	235	31	0.262	55	$t = \frac{5}{8}$ in.	
18.8(f)	7.2(f)	262	65	0.269	50	$t = \frac{3}{4}$ in.	
20.4(f)	7.8(f)	302	75	0.269		$t = 1$ in.	
12(f)		145	100			$t = \frac{1}{8}$ in.	
12(f)		190				$t = \frac{1}{8}\text{-}\frac{7}{8}$ in.	
						ELONG. 2 in., %	RED. Area, %
25	10.7	120	12	0.262	120	12.5	
25	10.7	130	16	0.262	120	20	
23	9.5	160	9–20(n)	0.26		18	
23	9.9	230	2–8(n)	0.26		6	
23	9.9	270	2–6(n)			5	
30	11.5	120	18(m)	0.284	55	30	50
30	11.5	130	23(m)	0.284	60	30	53
30	11.5	140	19(m)	0.284	65	26	40
30	11.5		30(m)	0.284			
30	11.5	160	22(m)	0.284	70	27	42
30	11.5	170	30(m)(u)	0.284		28(u)	68(u)
30	11.5	190	20(m)(u)	0.284	70	24	50
30	11.5	235	28(m)	0.284	60	18	42
30	11.5	269	25(m)	0.284		14(f)	30(f)
30	11.5	310		0.284		9(f)	22(f)
30	11.5	390	12(m)	0.284		8	15

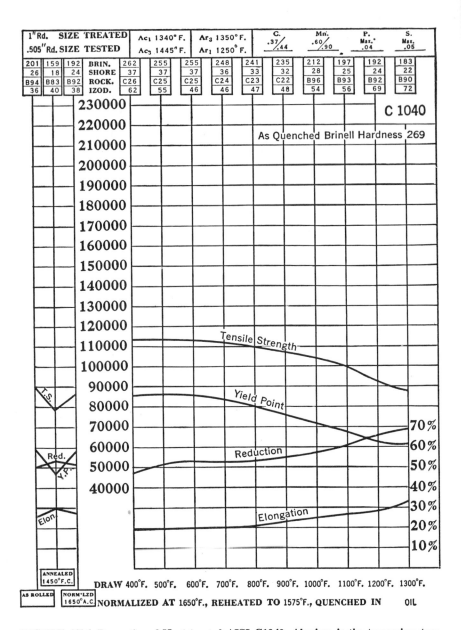

1"Rd. SIZE TREATED			Ac₁ 1340° F.	Ar₃ 1350° F.	C. .37/.44	Mn. .60/.90	P. Max. .04	S. Max. .05
.505"Rd. SIZE TESTED			Ac₃ 1445° F.	Ar₁ 1250° F.				

201	159	192	BRIN.	262	255	255	248	241	235	212	197	192	183
26	18	24	SHORE	37	37	37	36	33	32	28	25	24	22
B94	B83	B92	ROCK.	C26	C25	C25	C24	C23	C22	B96	B93	B92	B90
36	40	38	IZOD.	62	55	46	46	47	48	54	56	69	72

C 1040

As Quenched Brinell Hardness 269

230000
220000
210000
200000
190000
180000
170000
160000
150000
140000
130000
120000
110000
100000
90000
80000
70000
60000
50000
40000

Tensile Strength

Yield Point

Reduction

70%
60%
50%
40%
30%
20%
10%

Elongation

T.S.

Red.

Elon.

ANNEALED 1450°F.C.

AS ROLLED   NORM'LZD 1650°A.C.

DRAW 400°F. 500°F. 600°F. 700°F. 800°F. 900°F. 1000°F. 1100°F. 1200°F. 1300°F.

NORMALIZED AT 1650°F., REHEATED TO 1575°F., QUENCHED IN   OIL

FIGURE AF 1 Properties of Heat-treated AISI C1040. Abscissa is the tempering temperature. Average values. Such charts as these are guides to mechanical properties that may be expected when the diameter is, say, $\frac{1}{4}$ in. to $1\frac{1}{2}$ in. See Tables AT 8 and AT 9. Observe that the BHN increases with tensile strength and that $s_u \approx (500)(BHN)$—not so accurate at very low tempering temperatures. (Courtesy Bethlehem Steel Co., Bethlehem, Pa.).

.530″ Rd. SIZE TREATED			Ac₁ 1360°F.	Ar₃ 1265°F.	C. .38/.43	Mn. .70/.90	P. Max. .04	S Max. .04	Si. .20/.35	Ni. 1.10/1.40	Cr. .55/.75	Mo. Nil	Grain
.505″ Rd. SIZE TESTED			Ac₃ 1420°F.	Ar₁ 1200°F.									Size

HEAT TESTED — .39 .76 .013 .026 .25 1.20 .65 .08 6-8

187	285	BRIN.	555	514	477	461	388	352	311	285	262	223	
28	40	SHORE	75	70	65	63	54	49	44	40	37	32	
B91	C30	ROCK.	C55	C52	C49	C47	C41	C37	C33	C30	C26	C20	
49	20	IZOD.	35	22	22	24	40	51	66	80	97	105	

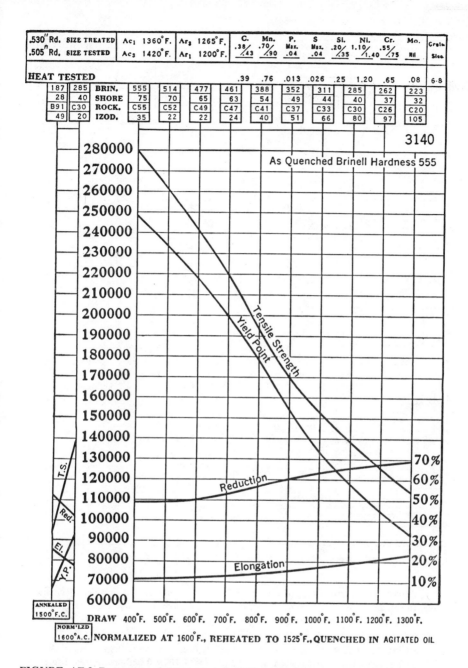

3140

As Quenched Brinell Hardness 555

Tensile Strength
Yield Point
Reduction
Elongation

T.S.
Red.
El.
Y.P.

70%
60%
50%
40%
30%
20%
10%

280000
270000
260000
250000
240000
230000
220000
210000
200000
190000
180000
170000
160000
150000
140000
130000
120000
110000
100000
90000
80000
70000
60000

ANNEALED 1500°F.C.

DRAW 400°F. 500°F. 600°F. 700°F. 800°F. 900°F. 1000°F. 1100°F. 1200°F. 1300°F.

NORM'LZD 1600°A.C. NORMALIZED AT 1600°F., REHEATED TO 1525°F., QUENCHED IN AGITATED OIL

FIGURE AF 2 Properties of Heat-treated AISI 3140. Single heat results. (Draw = Temper.) Notice the specified heat treatment and the size of the specimen. The ultimate strength $s_u \approx (500)(BHN)$ in psi. This material is widely used for heat-treated parts. For $R_c = 28$, decarburization of surface reduces endurance strength by 50%. For $R_c = 48$, decarburization of surface reduces endurance strength by 75%, from about $s'_n = 82$ ksi but the percentage is unusually high. See Fig. AF 3. (Courtesy Bethlehem Steel Co., Bethlehem, Pa.).

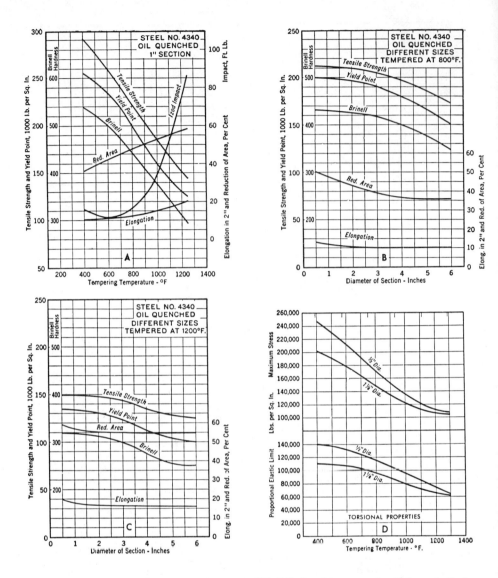

FIGURE AF 3 Properties of Heat-treated AISI 4340. Another manner in which mechanical properties might be depicted; charts A, B, and C are tensile properties; chart D gives torsional properties. An excellent general purpose alloy. Other miscellaneous endurance strengths of this steel follow. For $s_u \approx 270$ ksi:

Surface not decarburized, $s_n = 89$ ksi. Surface decarburized to 0.03 in., $s_n = 40$ ksi. Decarburized surface, shot peened, $s_n = 95$ ksi. OQT 1075, 0.625 dia., nitrided surface, $s_n = 120$ ksi. Test specimen from rolled stock, transversely, $s_n = 45$–70 ksi.

For $s_u \approx 160$ ksi; variation of endurance strengths with temperature.	Temp.	Reversed s_n', $R = -1$ (ksi)	Repeated, $R = 0$ (ksi)
	70°F	70	117
	600°F	61	96
	800°F	59	82
	1000°F	39	65

(Courtesy International Nickle Co., N.Y.)

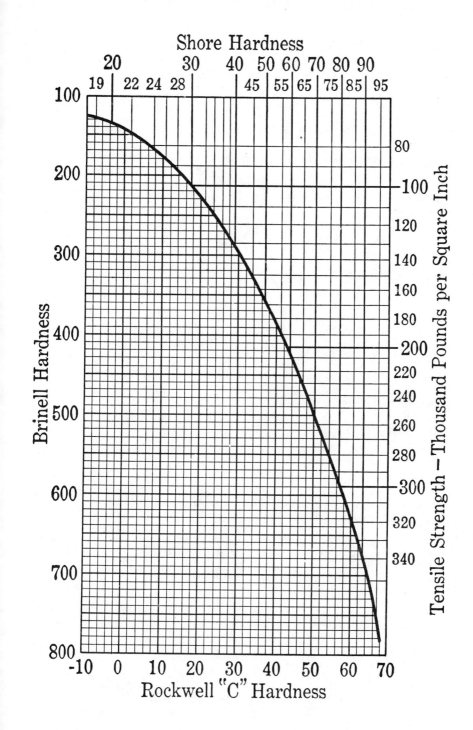

FIGURE AF 4 Relation of Hardness Numbers. (Courtesy International Nickle Co., N.Y.)

TABLE AT 7 TYPICAL PROPERTIES OF WROUGHT FERROUS METALS[1.1,2.1,2.3,2.6,2.10]

(See also charts for C1040, 3140, 4340; and Tables AT 8–AT 10, incl.).

For all wrought steels:

Modulus of elasticity in tension or compression, $E = 30 \times 10^6$ psi (For Wrought iron, $E = 28 \times 10^6$ psi).

Modulus of elasticity in shear or torsion, $G = 11.5 \times 10^6$ psi (For wrought iron, $G = 10 \times 10^6$ psi).

Yield strength in torsion (shear) frequently falls between $0.5s_y$ and $0.6s_y$. Use $0.6s_y$.

Endurance limit of a polished specimen, reversed bending, approximately $s_u/2$; Table AT 10.

Endurance limit in reversed torsion, approximately $0.6s_n'$; see § 4.7.

Poisson's ratio is about 0.25 to 0.33. Use 0.3 for steel.

MATERIAL AISI No.	CONDITION (c)	ULT. STR., ksi s_u	ULT. STR., ksi $s_s(d)$	TEN. YD. s_y, ksi	ELONG. 2 in. %	RED. Area %	BHN
Wrought Iron	As rolled	48(a)	36	25(a)	35		
Wrought Steel							
C1010 (k)	Cold drawn	67	50	55	25	57	137
C1015 (k)	Cold drawn	77	58	63	25	63	170
C1020	As rolled	65	49	48	36	59	143
C1020	Normalized	64	54*	50	39	69	131
C1020	Annealed	57	43	42	36.5	66	111
C1020 (k)	Cold drawn	78	58	66	20	59	156
C1022	As rolled	72	54	52	35	67	149
C1030	As rolled	80	60	51	32	56	179
C1035	As rolled	85	64	55	29	58	190
C1045	As rolled	96	72	59	22	45	215
C1095	Normalized	141	105	80	8	16	285
B1113 (k)	Cold finished	83	62	72	14	40	170
B1113	As rolled	70		45	25	40	138
C1118	As rolled	75	56	46	32	70	149
C1118 (k)	Cold drawn	80	60	75	16	57	180
C1144	OQT 1000	118	88	83	19	46	235
1340	OQT 1200	113	84	92	21	61	229
13B45	OQT 800	187	140	175	16	56	
2317(e)	OQT 1000	106	79	71	27	72	220
2340(e)	OQT 1000	137	103	120	22	60	285
3150	OQT 1000	151	113	130	16	54	300
3250(e)	QT 1000	166	122	146	16	52	340
4063	OQT 1000	180	135	160	14	43	375
4130	WQT 1100	127	95	114	18	62	260
4130(e)	Cold drawn	122	91	105	16	45	248
4340(e)	Cold drawn	122	91	105	15	45	248
4640(e)	OQT 1000	152	104	130	19	56	310
5140(e)	OQT 1000	150	113	128	19	55	300
5140(e)	Cold drawn	105	79	88	18	52	212
8630	Cold drawn 10%	115	86	100	22	53	222
8640	OQT 1000	160	120	150	16	55	330
8760	OQT 800	220	165	200	12	43	429
9255	OQT 1000	180	135	160	15	32	352
9440	OQT 1000	152	104	135	18	61	311
9850	OQT 1100	180	135	158	15	48	360

Density is about 0.284 lb/in.³ (0.28 lb/in.³ for wrought iron).

Coefficient of thermal expansion (linear) is 0.000 007 in./in.-°F (0.000 0006 5 for wrought iron). Varies significantly with large temperature change. See § 2.22 for cryogenic applications.

Notes: B preceding AISI No. indicates Bessemer, as B1113; C preceding indicates open hearth, as C1020. (a) Minimum values. (b) Annealed. (c) QT 1000 stands for "quenched and tempered at 1000°F," etc. (d) *Ultimate stress in shear* has been arbitrarily taken as (.75) (tensile ultimate); except starred * values which are test values. (e) 1-in. specimen. (f) Torsion. (g) Mill annealed. (h) Cold drawn. (i) See § 2.2 for definition. (j) Charpy V-notch, 70°F. (k) Properties depend on amount of cold work.

ROCK.	IZOD ft-lb.	MACHIN-ABILITY (*i*)	SOME TYPICAL USES; REMARKS
B60		50(h)	ASTM A85-36, A41-36
		50	Bars, strips, sheet, plate. Cold drawn shapes.
	137(j)	50	Bars, sheets. Table AT 8. For carburizing: Table AT 11.
B79	64	64	Structural steel; plate, sheet, strip, wire.
B74	72		Carburizing grade, Table AT 11.
B66	80		General purpose.
B83		62	Misc. machine parts are cold forged; bars.
B81	60	70(h)	General purpose.
B88	55	60	Machinery parts. Table AT 8.
B91	45	57	Machine parts. May be heat treated. Table AT 9.
B96	30	51	Large shafts.
C25	3	39	Tools, springs. Usually heat treated. Table AT 9.
B87		135	Free cutting; high sulfur.
B76			Free cutting; high sulfur.
B81	80	82	Free cutting; not usually welded. Carburized, Table AT 11.
	110(j)	85	Table AT 8 for C1117.
C22	36	65(b)	Free cutting. High sulfur. Tables AT 8 and AT 9 for C1137.
C31	95	45(g)	(1.75% Mn). Manganese steel.
C42			1345 with boron for improved hardenability.
B97	85	55(h)	($3\frac{1}{2}$% Ni)—Gears, pump liners, etc.
C30	50	31	($3\frac{1}{2}$% Ni)—Gears, etc.
C32	46		(1.25% Ni, 0.8% Cr.) Gears, bolts, shafts, etc.
C36.5	30	55(b)	(1.85% Ni, 1.05% Cr) Gears, etc.
C40	59		(0.25% Mo) Shafts, bars, etc.
C25	85	65(b)	(0.95% Cr, 0.20% Mo) Shafts, forgings, pins, aircraft tubes.
		45(g)	(1.85% Ni, 0.8% Cr, 0.25% Mo) General purpose. Fig. AF3.
C33	41	55(b)	(1.85% Ni, 0.25% Mo).
C32		60(b)	(0.80% Cr) Gears, shafts, pins, etc.
		60(g)	
			(0.55% Ni, 0.5% Cr, 0.2% Mo). Table AT 9.
C35	36	60(b)	(0.55% Ni, 0.50% Cr, 0.20% Mo).
C46	19	50(b)	(0.55)% Ni, 0.50% Cr, 0.25% Mo) Tools, gears, bolts.
C36	7	45(b)	(2.00% Si, 0.82% Mn) Springs, chisels, tools.
C33	73	60(b)	(0.45% Ni, 0.4% Cr, 0.11% Mo).
C37	50	50(b)	(1% Ni, 0.8% Mn, 0.8% Cr. 0.25% Mo) Heavy duty; general.

TABLE AT 8 TYPICAL PROPERTIES OF STEEL— VARIOUS SIZES AND CONDITIONS[2.6]

(a) Turned. (b) 10%. (c) Inconsistent—from different mill.

AISI NO.	CONDITION	ROD DIA. in.	ULT. STR. s_u ksi	TEN YD. s_y ksi	ELONG. 2 in., %	RED. Area, %	BHN	IZOD ft-lb.
C1015	As rolled	$\frac{1}{2}$	61	45.5	39	61	126	81
	Annealed	1	56	42	37	69.7	111	83
	Normalized	$\frac{1}{2}$	63	48	38.6	71	126	85
	Normalized	1	61.5	47	37	69.6	121	85
	Normalized	2	60	44.5	37.5	69.2	116	86
	Normalized	4	59	41.8	36.5	67.8	116	83
C1117	As rolled	$\frac{1}{2}$	70.6	44.3	33	63	143	60
	Annealed	1	62	40.5	32.8	58	121	69
	Normalized	$\frac{1}{2}$	69.7	45	34.3	61	143	70
	Normalized	2	67	41.5	33.5	64.7	137	83
	Normalized	4	63.7	35	34.3	64.7	126	84
C1030	As rolled	$\frac{1}{2}$	80	51	32	54	179	55
	Annealed	1	67	49	31	57.9	126	51
	Normalized	$\frac{1}{2}$	77.5	50	32	61.1	156	69
	Normalized	4	72.5	47	29.7	56.2	137	61
	WQT 1000	1	88	68	28	68.6	179	92
C1137	As rolled	$\frac{1}{2}$	93	55	26	63	192	61
	Annealed	1	85	50	27	54	174	37
	Normalized	$\frac{1}{2}$	98	58	25	58	201	69
	Normalized	2	96	49	22	51	197	21
	Cold drawn	1	103	93	15	56	217	
C1045	Annealed	1	90	55	27	54	174	32(c)
	Normalized	1	99	61	25	49	207	48(c)
	Hot rolled(a)	1	87	54	27	56	187	51(c)
	Cold drawn(b)	2	100	85	19	45	235	
	WQT 1000	$\frac{1}{2}$	130	110	16	56	260	75(c)
	WQT 1200	$\frac{1}{2}$	110	84	23	61	220	
	WQT 1000	2	110	70	23	50	205	85(c)
	WQT 1200	2	98	64	26	58	190	
	WQT 1000	4	94	59	25	49	180	62(c)
	WQT 1200	4	93	55	28	55	186	
C1050	As rolled	$\frac{1}{2}$	102	58	18	37	229	23
	Annealed	1	92	53	23.7	40	187	12
	Normalized	$\frac{1}{2}$	111	62	21.5	45	223	17
	Normalized	4	100	56	21.7	41.6	201	20
	Cold drawn	1	113	95	12	35	229	
	OQT 1100	$\frac{1}{2}$	122	81	22.8	58	248	22
	WQT 1100	$\frac{1}{2}$	119	88	21.7	60	241	51
	OQT 1100	2	112	68	23	55.6	223	20
	WQT 1100	2	117	78.5	23	61	235	24
	OQT 1100	4	101	58.5	25	54.5	207	21
	WQT 1100	4	112	68	23.7	55.5	229	15

TABLE AT 9 TYPICAL PROPERTIES OF
HEAT-TREATED STEELS[2.3,2.6]

Values in this table have been taken from charts such as those of Figs. AF 1–AF 3. To get the strength or Brinell number for any other tempering (drawing) temperature, make a straight line interpolation between the values given. Extrapolation to lower temperatures might sometimes give a reasonable estimate, but extrapolation should not be relied upon. (a) Do not interpolate using this value.

AISI NO. (Quenching medium)	SIZE	TEM- PERED AT, °F	ULT. STR. s_u ksi	TEN. YD. s_y ksi	BHN	ELONG. 2 in. %	IZOD, ft-lb.
C1035	1″	600	118	87	240	11	40
(water)	1″	1000	102	73	200	22	57
	1″	1300	85	57	170	29	93
C1095	$\frac{1}{2}$″	800	176	112	363	11	6
(oil)	$\frac{1}{2}$″	1100	145	88	293	17	6
	4″	1100	130	65	262	17	5
C1137	$\frac{1}{2}$″	700	135	115	277	12	13(a)
(oil)	$\frac{1}{2}$″	1000	111	88	229	23	61
	2″	1000	105	63	217	23	31
2330	$\frac{1}{2}$″	600	210	195	429	13	39
Nickel	$\frac{1}{2}$″	1000	135	126	277	20	77
Steel	$\frac{1}{2}$″	1300	107	91	217	26	109
(water)	4″	1000	105	85	207	26	87
4140	$\frac{1}{2}$″	500	270	241	534	11	8(a)
Cr-Mo	$\frac{1}{2}$″	800	210	195	429	15	21
(oil)	$\frac{1}{2}$″	1200	130	115	277	21	83
	4″	1200	112	83	229	23	87
4150	$\frac{1}{2}$″	800	228	215	444	10	12(a)
Cr-Mo (oil)	$\frac{1}{2}$″	1200	159	141	331	16	53(a)
5150	$\frac{1}{2}$″	800	210	195	415	11	17(a)
Chromium	$\frac{1}{2}$″	1000	160	149	321	15	39
(oil)	$\frac{1}{2}$″	1200	127	117	269	21	59
6152	$\frac{1}{2}$″	700	246	224	495	10	9(a)
Cr-V	$\frac{1}{2}$″	1000	184	173	375	12	30
(oil)	$\frac{1}{2}$″	1200	142	131	293	18	65
	2″	1200	121	94	241	21	45(a)
8630	$\frac{1}{2}$″	800	185	174	375	14	58
Ni-Cr-Mo	$\frac{1}{2}$″	1100	137	125	285	20	95
(water)	4″	1100	96	72	197	25	104
8742	1″	700	226	203	455	11	14(a)
Ni-Cr-Mo	1″	1200	130	110	262	21	67(a)
(oil)	4″	1200	118	91	235	22	
9261	$\frac{1}{2}$″	800	259	228	514	10	12
Si-Mn	$\frac{1}{2}$″	900	215	192	429	11	13
(oil)	$\frac{1}{2}$″	1200	147	124	311	17	35(a)
9840	1″	700	237	214	470	11	10(a)
Ni-Cr-Mo	1″	1200	140	120	280	19	65(a)
(oil)	6″	1000	151	131	302	16	

TABLE AT 10 MISCELLANEOUS ENDURANCE LIMITS AND ENDURANCE STRENGTHS[2.1,2.3,2.5,2.9,2.12,2.16]

Specimens 0.5 in. or smaller. See also Tables AT 3, AT 4, AT 6, AT 7, Fig. AT 3. Endurance ratio s_n/s_u decreases as size of section increases, to as low as 0.35 for 6 in. dimension in cast steel.

Notes: (a) Manganese steel. (b) Number of cycles is indefinitely large unless specified. Cy. = cycles. (c) By analogy (not a test value). (d) Depends on the number of cycles. (e) Permanent mold.

MATERIAL	CONDITION	s_n KSI AT NO. OF CY. (b)	$\dfrac{s_n}{s_u}$ (d)	s_y KSI	$\dfrac{s_y}{s_n}$
Wrought iron	Longitudinal	23	0.49	28	1.22
Wrought iron	Transverse	19	0.55	25(c)	1.31
Cast iron	ASTM 30	12	0.38		
Cast iron	ASTM 30	16 at 10^5			
Cast iron	ASTM 30	21 at 10^4			
Cast steel, 0.18 % C	As cast	31.5	0.45	36	1.14
Cast steel, 0.18 % C	Cast and annealed	34.5	0.45	37	1.07
Cast steel, 0.25 % C	Cast and normalized	35	0.46	45	1.29
Cast steel, 1330(a)	Cast, N&T 1200	48	0.49	61	1.27
Cast steel, 1330(a)	Cast, WQT (269 BHN)	58	0.48	106	1.83
Cast steel, 4340	Cast, WQT 1100	64	0.40	148	2.32
Cast steel, 8630	Cast, N&T 1200	54	0.49	85	1.57
Cast steel, 8630	Cast, WQT (286 BHN)	65	0.47	125	1.92
Wrought Steel 1015					
1015	Cold drawn (10 % work)	40	0.57	63	1.58
1020	As rolled	45 at 10^4		48	1.08
1020	As rolled	40 at 10^5		48	1.20
1020	As rolled	33 at 10^6		48	1.45
1035	Cold drawn	46(c)	0.50	78	1.69
1035	In air	40.6	0.46	58	1.43
1035	In brine	24.6		58	2.36
1035	In sulfur	10.6		58	5.48
1040	Cold drawn (10 % work)	54	0.54	85	1.57
1040	Cold drawn (20 % work)	59	0.5	92	1.56
1117	Cold drawn	40(c)	0.50(c)	68	1.70
1141	Cold drawn	50	0.46	90	1.8
13B45	OQT 1100	68	0.54	112	1.65
1144	Elevated temp. drawn (ETD)	72	0.48	140	1.95
2317	In air	52	0.61	50	0.96
2317	In brine	31.6		50	1.58
2317	In sulfur	23.9		50	2.09
2320	Hot rolled rod	48	0.50	51	1.06
2320	Carburized, case hardened	90	0.53	140	1.56
3120	Carburized, case hardened	90	0.64	100	1.11
4340	At 1000°F (OQT 1150)	40			
6150	Heat treated	96	0.46	190	1.98
8630	Cold drawn (20 %)	62	0.51	107	1.73
94B40	OQT 1100	70	0.51	119	1.70
Nitralloy N	Nitrided	124	0.65	180	1.45
Nitralloy 135, modified	Un-nitrided	45			
Nitralloy 135, modified	Nitrided	90	0.66	140	1.56
Nitralloy 135, modified	Notched and un-nitrided	24			
Nitralloy 135, modified	Notched and nitrided	80	0.59	140	1.75
Stainless steel 316	Annealed bar	38	0.37	35	0.92

MATERIAL	CONDITION	S_n KSI AT NO. OF CY. (b)	$\dfrac{S_n}{S_u}$ (d)	S_y KSI	$\dfrac{S_y}{S_n}$
Stainless steel 403 .	Annealed bar	40	0.57	37	0.67
Stainless steel 403 .	Bars, heat treat. to $R_B = 97$	55	0.50	85	1.54
Stainless steel 410 .	Bars, OQT to $R_B = 97$	58	0.52	85	1.47
Stainless steel 410 .	ditto, except at 850°F	43			
Stainless steel 418 .	OQT 1200	75	0.54	108	1.43
Stainless steel 430 .	Annealed and cold drawn; 185 BHN	46	0.61	50	1.09
Aluminum 2011 . .	Wrought, T3	18 at 5×10^8	0.33	43	2.39
Aluminum 2014 . .	Wrought, T4	20 at 5×10^8	0.32	42	2.10
Aluminum 2014 . .	Wrought, T6	18 at 5×10^8	0.26	60	3.33
Aluminum 2014 . .	Ditto	30 at 10^6	0.43	60	2.00
Aluminum 2014 . .	Ditto, 500°F	5 at 5×10^8	0.45	8.5	1.70
Aluminum 2014 . .	T6, reversed axial	15 at 5×10^8	0.21	60	4.00
Aluminum 5052 . .	Cold worked, H32	17 at 5×10^8	0.51	28	1.65
Aluminum 5052 . .	Cold worked, H36	19 at 5×10^8	0.47	35	1.84
Aluminum 6063 . .	Wrought, T5	10 at 5×10^8	0.37	21	2.1
Aluminum 7079 . .	Wrought, T6	23 at 5×10^8	0.30	68	2.96
Aluminum, 142 alloy .	Sand casting, T77	10.5 at 5×10^8	0.35	23	2.19
Aluminum, 142 alloy .	Casting, T61(e)	9.5 at 5×10^8	0.20	42	4.42
Alum. bronze (10%)	Extruded, heat treated	34 at 7×10^7	0.44	50(c)	1.47
Alum. bronze (10%)	Sand cast, annealed	28 at 8×10^7	0.34	40	1.43
Cartridge brass (70-30)	0.08″ spring wire	22 at 10^8	0.17	65(c)	2.96
Cartridge brass (70-30)	Half hard, 1″ rod	22 at 5×10^7	0.31	52	2.36
Free-cutting brass .	Half hard, 2″ rod., SAE 72	14 at 3×10^8	0.25	44	3.14
Commercial bronze	0.08″ hard wire	23 at 10^8	0.31	60(c)	2.61
Leaded tin bronze .	Sand cast, alloy 2A(Navy M)	11 at 10^8	0.29	18	1.64
Low brass (80-20) .	Spring hard, 0.04″ strip	24 at 2×10^7	0.26	65	2.70
Low brass (80-20) .	0.08″ spring wire	26 at 10^8	0.21	88(c)	3.38
Manganese bronze .	Sand cast, alloy 8A	21.2 at 10^8	0.30	28	1.32
Manganese bronze .	Sand cast, alloy 8C	25 at 10^8	0.24	70	2.8
Silicon bronze, type A	Half-hard rod	30 at 3×10^8	0.39	45	1.50
Silicon bronze, type B	Hot rolled	19 at 5×10^7			
Silicon bronze, type B	Extruded	20 at 5×10^7	0.29	55(c)	2.75
Silicon bronze, type B	Cold drawn, 72% reduction	30 at 3×10^8	0.32	69(c)	2.30
Silicon bronze, type B	0.08″ hard wire	25 at 10^8	0.28	67	2.68
Magnesium (AZ63A).	Cast, T5	11 at 5×10^8	0.38	15	1.36
Magensium (AZ31B).	Extruded bar	15 at 5×10^8	0.41	22	1.47
Inconel (Ni-Cr) . .	Cold drawn	40 at 10^8	0.38	80	2.00
Inconel . . .	As forged or hot rolled	38 at 10^8	0.42	35	0.92
Monel (67 Ni, 30 Cu)	Annealed rod	31 at 10^8	0.41	30	0.97
Monel . . .	Cold drawn rod	42 at 10^8	0.42	75	1.78
Monel . . .	Annealed. In brackish water	21 at 10^8	0.28	30	1.43
K-Monel (3 Al) .	Cold drawn, age hardened	45 at 10^8	0.30	110	2.44
Titanium (5 Al, 2.5 Sn)	Formed; ground finish	60	0.5	110	1.83

TABLE AT 11 TYPICAL CORE PROPERTIES OF CARBURIZED STEELS[2.3]

Carburizing is done at about 1700°F. A tempering temperature of 300°F produces maximum case hardness; 450°F results in improved impact strength.

Notes. (a) Nominal size of specimen, 1 in. (b) ½-in. specimen. (c) 2-in. specimen. (d) 4-in. specimen. (e) Abbreviations: "SOQT 450," single oil quench and temper at 450°F; "DWQT 300," double water quench and temper at 300°F; Q, quench; P, pot. (f) Of the order of other hardnesses shown. (g) Case thickness depends on temperature and time of carburizing; for example, at 1700°F for 4 hr., the case should be of the order of 0.05 in.; at 1700°F for 8 hr., about 0.06 in. As seen from the values given, these are not hard and fast rules.

AISI NO.	CONDITION (e)	Ult. Str. s_u ksi	Ten. Yd. s_y ksi	Elon. % 2"	Red. Area %	BHN	Izod ft-lb.	Rock. Hard. R_C	Thick. in. (8hr.)
C1015(b)	SWQT 350	73	46	32	71	149	91	C62	0.048
C1020(a)	DWQT 300	85	55	33	65	170		(f)	(g)
C1020(a)	SWQT 300	80	50	30	60	160		(f)	(g)
C1117(b)	SWQT 350	96	59	23	53	192	33	C65	0.045
2115(a)	DO(or W)QT 300	90	60	30	70	185	70	(f)	(g)
2317(a)	DOQT 300	95	60	35	65	195	85	(f)	(g)
2317(a)	DWQT 300	100	65	30	60	210	70	(f)	(g)
2515(a)	DOQT 300	170	130	14	50	352	40	(f)	(g)
3115(a)	DOQT 300	100	70	25	55	212	55	(f)	(g)
3215(a)	SOQT 300	141	110	17	50		45	(f)	(g)
E3310(b)	SOQT 450	180	149	14.5	58	363	57	C57.5	0.047
E3310(b)	DOQT 300	177	143	15.3	58	352	47	C61	0.047
3415(a)	SOQT 300	130	95	18	52	285	55	(f)	(g)
3415(a)	DOQT 300	135	105	19	55	300	60	(f)	(g)
4320(b)	Direct OQ from P 300	217	159	13	50	429	32	C60.5	0.060
4320(b)	DOQT 450	145	94	21.8	56	293	48	C59	0.075
4620(b)	DOQT 300	122	77	22	56	248	64	C62	0.060
4620(b)	DOQT 450	115	77	22.5	62	235	78	C59	0.060
4820(b)	SOQT 300	207	167	13.8	52	415	44	C61	0.047
4820(b)	SOQT 450	205	184	13	53	415	47	C57.5	0.047
8620(b)	SOQT 300	188	149	11.5	51	388	26	C64	0.075
8620(b)	SOQT 450	167	120	14.3	53	341	29	C61	0.076
8620(b)	DOQT 300	133	83	20	56	269	55	C64	0.070
E9310(b)	Direct OQ from P 300	179	144	15.3	59	375	57	C59.5	0.039
E9310(b)	SOQT 300	173	135	15.5	60	363	61	C62	0.047
E9310(b)	DOQT 300	174	139	15.3	62	363	54	C60.5	0.055
E9310(a)	SOQT 300	159	122	15.5	57	321	68	(f)	(g)
E9310(c)	SOQT 300	145	108	18.5	66	293	93	(f)	(g)
E9310(d)	SOQT 300	136	94	19	62	277	93	(f)	(g)

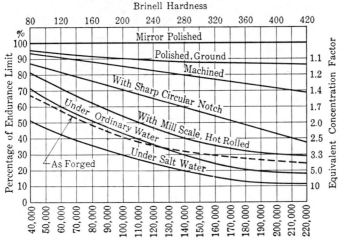

FIGURE AF 5 Reduction of Endurance Strength of Steel.[4.4] Some effects are stress raisers, some are strength reducers. See § 3.14 for the approximate roughnesses of the corresponding surfaces. A ground surface is not expected to have a roughness greater than 100 microinches. The curve for "as forged," adapted from Lipson, et al.[4.2], assumes decarburization of the surface. The machined surface is a good one, light cut, fine feed. Polishing leaves a residual compressive stress, helpful against fatigue.

TABLE AT 12 VALUES OF K_f FOR SCREW THREADS [4.2] For tension or bending. Not K_t.

| KIND OF THREAD | ANNEALED | | HARDENED | |
	Rolled	Cut	Rolled	Cut
Sellers, Amer. Nat'l., Sq. Th.	2.2	2.8	3.0	3.8
Whitworth Rounded Roots	1.4	1.8	2.6	3.3
Dardelet	1.8	2.3	2.6	3.3

TABLE AT 13 VALUES OF K_f FOR KEYWAYS.[4.2] See § 10.4

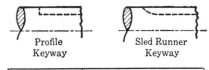

| | Profile Keyway | | Sled Runner Keyway | |

| KIND OF KEYWAY | ANNEALED | | HARDENED | |
	Bend-ing	Tor-sion	Bend-ing	Tor-sion
Profile	1.6	1.3	2.0	1.6
Sled-runner	1.3	1.3	1.6	1.6

Semicircular

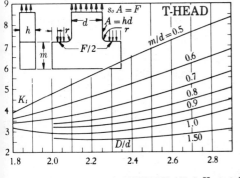

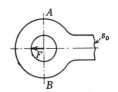

Approx. $K_t = 2.8$. $s_{max} = K_t s_0$ at AB inside hole.

INTERFERENCE FITS

$K_f = 1.5$ to 4, Examples: Cold-rolled shaft $K_f = 1.9$. Heat-treated shaft $K_f = 2.6.7$-in as forged $K_f = 3$.

FIGURE AF 6 K_t and K_f for T-head and Miscellaneous.[4.2]

125

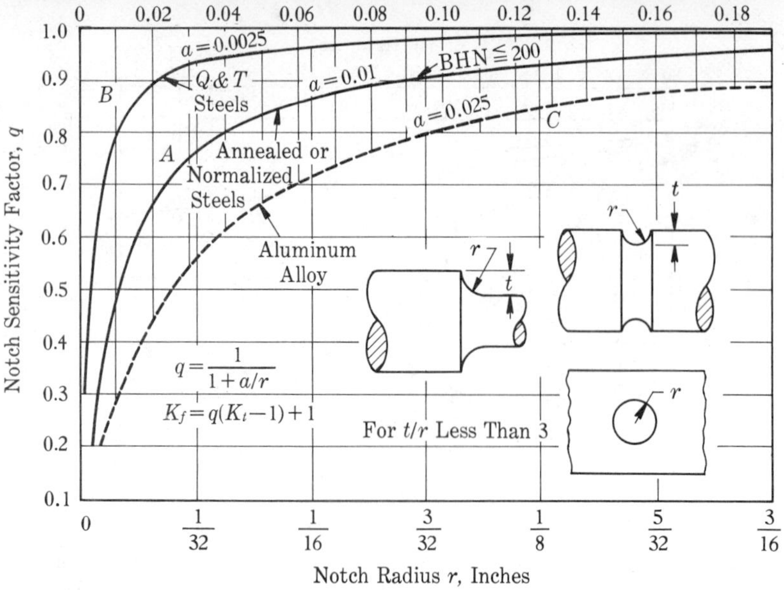

Notch Radius r, Inches
0 0.02 0.04 0.06 0.08 0.10 0.12 0.14 0.16 0.18
Notch Sensitivity Factor, q
1.0 0.9 0.8 0.7 0.6 0.5 0.4 0.3 0.2 0.1
a = 0.0025
Q & T Steels
B
a = 0.01
BHN ≦ 200
A
Annealed or Normalized Steels
a = 0.025
C
Aluminum Alloy
q = 1/(1 + a/r)
K_f = q(K_t − 1) + 1
For t/r Less Than 3
r
t
0 1/32 1/16 3/32 1/8 5/32 3/16
Notch Radius r, Inches

FIGURE AF 7 Average Notch Sensitivity Curves. Applicable particularly to normal stresses; used also for shear stresses. (After R. E. Peterson)[4.1.4.57]

FIGURE AF 8 Flat Plate with Central Hole—Tension and Bending.[4.2,4.16,4.21] Use solid curve for bending in plane of paper; and for rod ($D = h$) with a hole (d) in tension. Symmetric loading. (Some evidence[4.61] that values for a tensile rectangular body are overly conservative.)

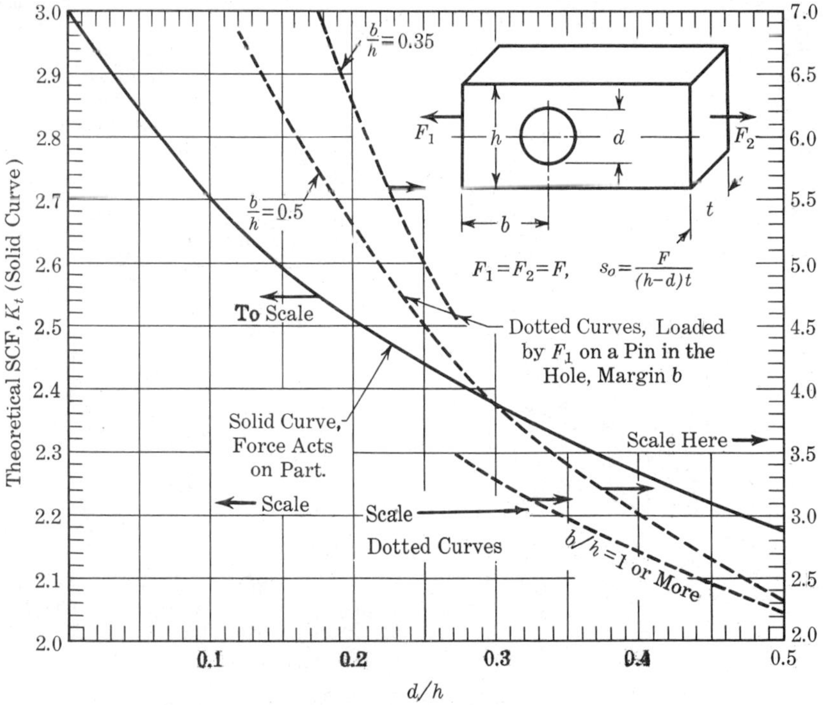

Theoretical SCF, K_t (Solid Curve)
3.0 2.9 2.8 2.7 2.6 2.5 2.4 2.3 2.2 2.1 2.0
Theoretical SCF, K_t (Dotted Curves)
7.0 6.5 6.0 5.5 5.0 4.5 4.0 3.5 3.0 2.5 2.0
b/h = 0.35
b/h = 0.5
F_1
F_2
h
d
b
t
F_1 = F_2 = F, s_o = F/((h − d)t)
To Scale
Dotted Curves, Loaded by F_1 on a Pin in the Hole, Margin b
Solid Curve, Force Acts on Part.
Scale Here
Scale
Scale Dotted Curves
b/h = 1 or More
0.1 0.2 0.3 0.4 0.5
d/h

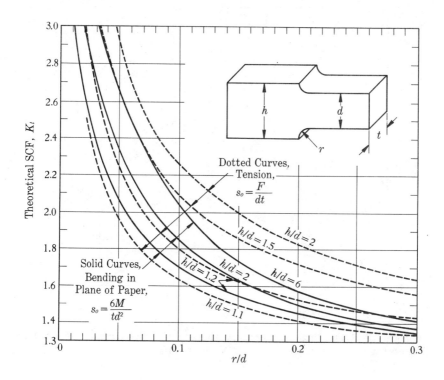

FIGURE AF 9 Flat Plate with Fillets. The tensile load is central. For $h/d = 1.1$, tensile and bending curves are very close together down to $r/d = 0.04$. (After R. E. Peterson)[4.21]

FIGURE AF 10 Flat Plate with Grooves.

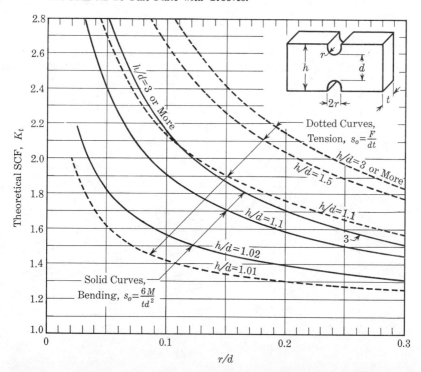

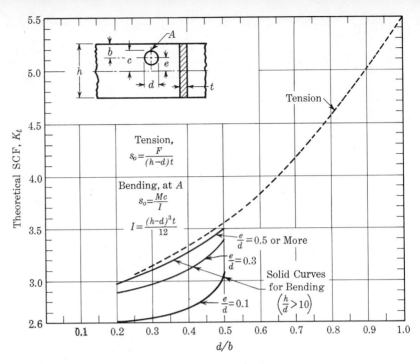

FIGURE AF 11 Flat Plate with Eccentric Hole.[4.2,4.20] For tension, if $h < 20d$, K_t is somewhat smaller than shown. (After R. E. Peterson).[4.21]

FIGURE AF 12 Shaft with Fillet. The tensile load is central. Torsion curve $D/d = 1.2$ approximates the bending curve for $D/d = 1.01$; torsion curve $D/d = 2$, approximates the bending curve for $D/d = 1.02$ (down to $r/d \sim 0.04$). Bending curve $D/d = 1.1$ approximates the tensile curve $D/d = 1.1$. (After R. E. Peterson).[4.21]

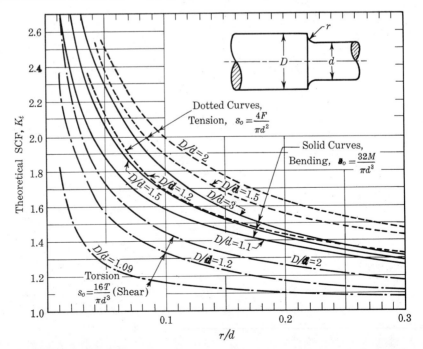

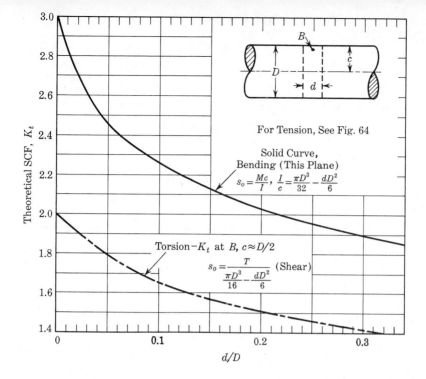

FIGURE AF 13 Shaft with Radial Hole—Bending and Torsion. Maximum torsional stress for this case falls slightly inside the shaft diameter on the inside surface of the hole at some point B. (After R. E. Peterson).[4.21]

FIGURE AF 14 Shaft with Groove. Use the solid curve for $D/d = 1.01$ for tension as well as bending (approximate).

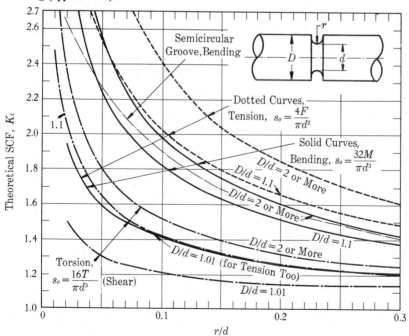

Selected values abstracted from ASA B1.1-1960,[5.1] which should be referred to for details on proportions and tolerances, and for other thread series; dimensions for a class 3 fit, external thread. The minor diameter for the internal thread is not quite the same as for the external thread. The *tensile stress area* is that area corresponding to a diameter that is approximately the average of the pitch and minor diameters; detail in Ref. *(5.1)*. Selected values of 12-thread series above $1\frac{1}{2}$-in. size.

SIZE	BASIC MAJOR DIA. in.	COARSE (UNC) Th./In.	Minor Dia. Ext. Th.	Stress Area, A_s sq. in.	FINE (UNF), AND 12 UN Th./In.	Minor Dia. Ext. Th.	Stress Area, A_s sq. in.	WIDTH ACROSS FLATS, REGULAR, UNFINISHED; A in. (Selected from ASA B 18.2-1952.) Head	Nut
0	0.0600				80	0.0447	0.00180	For square	
1	0.0730	64	0.0538	0.00263	72	0.0560	0.00278	heads and	
2	0.0860	56	0.0641	0.00370	64	0.0668	0.00394	nuts, A is diff-	
3	0.0990	48	0.0734	0.00487	56	0.0771	0.00523	erent in the	
4	0.1120	40	0.0813	0.00604	48	0.0864	0.00661	smaller sizes (below 7/8)	
5	0.1250	40	0.0943	0.00796	44	0.0971	0.0083	from these	
6	0.1380	32	0.0997	0.00909	40	0.1073	0.01015	values. See	
8	0.1640	32	0.1257	0.0140	36·	0.1299	0.01474	detail in the	
10	0.1900	24	0.1389	0.0175	32	0.1517	0.0200	standard.	
12	0.2160	24	0.1649	0.0242	28	0.1722	0.0258		
$\frac{1}{4}$	0.2500	20	0.1887	0.0318	28	0.2062	0.0364	$\frac{7}{16}$	$\frac{7}{16}$
$\frac{5}{16}$	0.3125	18	0.2443	0.0524	24	0.2614	0.0580	$\frac{1}{2}$	$\frac{1}{2}$
$\frac{3}{8}$	0.3750	16	0.2983	0.0775	24	0.3239	0.0878	$\frac{9}{16}$	$\frac{9}{16}$
$\frac{7}{16}$	0.4375	14	0.3499	0.1063	20	0.3762	0.1187	$\frac{5}{8}$	$\frac{11}{16}$
$\frac{1}{2}$	0.5000	13	0.4056	0.1419	20	0.4387	0.1599	$\frac{3}{4}$	$\frac{3}{4}$
$\frac{9}{16}$	0.5625	12	0.4603	0.182	18	0.4943	0.203	$\frac{13}{16}$	$\frac{7}{8}$
$\frac{5}{8}$	0.6250	11	0.5135	0.226	18	0.5568	0.256	$\frac{15}{16}$	$\frac{15}{16}$
$\frac{3}{4}$	0.7500	10	0.6273	0.334	16	0.6733	0.373	$1\frac{1}{8}$	$1\frac{1}{8}$
$\frac{7}{8}$	0.875	9	0.7387	0.462	14	0.7874	0.509	$1\frac{5}{16}$	$1\frac{5}{16}$
1	1.0000	8	0.8466	0.606	12	0.8978	0.663	$1\frac{1}{2}$	$1\frac{1}{2}$
$1\frac{1}{8}$	1.125	7	0.9497	0.763	12	1.0228	0.856	$1\frac{11}{16}$	$1\frac{11}{16}$
$1\frac{1}{4}$	1.2500	7	1.0747	0.969	12	1.1478	1.073	$1\frac{7}{8}$	$1\frac{7}{8}$
$1\frac{3}{8}$	1.375	6	1.1705	1.155	12	1.2728	1.315	$2\frac{1}{16}$	$2\frac{1}{16}$
$1\frac{1}{2}$	1.5000	6	1.2955	1.405	12	1.3978	1.581	$2\frac{1}{4}$	$2\frac{1}{4}$
$1\frac{3}{4}$	1.7500	5	1.5046	1.90	12	1.6478	2.1853	$2\frac{5}{8}$	$2\frac{5}{8}$
2	2.0000	$4\frac{1}{2}$	1.7274	2.50	12	1.8978	2.8892	3	3
$2\frac{1}{4}$	2.2500	$4\frac{1}{2}$	1.9774	3.25	12	2.1478	3.6914	$3\frac{3}{8}$	$3\frac{3}{8}$
$2\frac{1}{2}$	2.5000	4	2.1933	4.00	12	2.3978	4.5916	$3\frac{3}{4}$	$3\frac{3}{4}$
$2\frac{3}{4}$	2.7500	4	2.4433	4.93	12	2.6478	5.5900f	$4\frac{1}{8}$	$4\frac{1}{8}$
3	3.0000	4	2.6933	5.97	12	2.8978	6.6865	$4\frac{1}{2}$	$4\frac{1}{2}$
$3\frac{1}{4}$	3.2500	4	2.9433	7.10	12	3.1478	7.8812	$4\frac{7}{8}$	
$3\frac{1}{2}$	3.5000	4	3.1933	8.33	12	3.3978	9.1740	$5\frac{1}{4}$	
$3\frac{3}{4}$	3.7500	4	3.4433	9.66	12	3.6478	10.5649	$5\frac{5}{8}$	
4	4.0000	4	3.6933	11.08	12	3.8978	12.0540	6	

TABLE AT 15 NOMINAL DIMENSIONS OF VARIOUS GAGES

The Washburn and Moen (W & M) gage, called also the steel-wire gage, is used for steel wire. The American Wire or Brown and Sharpe (B & S) gage is used for monel, bronze, copper, aluminum, and brass wires. Standard wire sizes other than those in the table include multiples of $\frac{1}{32}$ up to $\frac{9}{16}$ in. The tendency is to specify the decimal size of the wire. There is also a music-wire gage. And much smaller sizes than those listed are available.

| GAGE NO. | WIRE DIAMETER, IN. | | PLATE THICKNESS, IN. |
	W & M Ferrous	B & S Nonferrous	U.S. Standard
7–0	0.4900		0.500
6–0	0.4615		0.469
5–0	0.4305		0.438
4–0	0.3938	0.460	0.406
3–0	0.3625	0.401	0.375
2–0	0.3310	0.365	0.344
0	0.3065	0.325	0.313
1	0.2830	0.289	0.281
2	0.2625	0.258	0.266
3	0.2437	0.229	0.250
4	0.2253	0.204	0.234
5	0.2070	0.182	0.219
6	0.1920	0.162	0.203
7	0.1770	0.144	0.188
8	0.1620	0.128	0.172
9	0.1483	0.114	0.156
10	0.1350	0.102	0.141
11	0.1205	0.091	0.125
12	0.1055	0.081	0.109
13	0.0915	0.072	0.094
14	0.0800	0.065	0.078
15	0.0720	0.057	0.070
16	0.0625	0.051	0.063
17	0.0540	0.045	0.056
18	0.0475	0.040	0.050

TABLE AT 16 APPROXIMATE FREE LENGTHS AND SOLID HEIGHTS

(P = pitch of coils, N_c = number of *active* coils, D_w = diameter of wire)

TYPE OF ENDS	FREE LENGTH	TOTAL COILS	SOLID HEIGHT
Plain	$PN_c + D_w$	N_c	$D_w N_c + D_w$
Plain ground	PN_c	N_c	$D_w N_c$
Squared	$PN_c + 3D_w$	$N_c + 2$	$D_w N_c + 3D_w$
Squared and ground	$PN_c + 2D_w$	$N_c + 2$	$D_w N_c + 2D_w$

TABLE AT 17 MECHANICAL PROPERTIES OF WIRE FOR COIL SPRINGS

For extension springs, use 0.8 times the value in column (5) for the maximum occasional stress.

The stress is $s = Q/D_w^x$ wherever this form appears; to be used with equation (6.1); always include the curvature factor except for the mean stress in fatigue design. The values given apply when the spring is not preset and not peened, except as stated. For live loads, stress values for the *steels* may be increased by 25% for shot-peened coils. For preset *steel* springs, the static stress and "solid stress" may be some 40–50% greater than given by columns (3) and (5). See notes (f) and (n) below. The computed design stresses are not to be interpreted as being exact values. Reduce design stresses 50% for shock loads (analogous to hammer blows). Where a maximum stress is given, use it for wire sizes smaller than the specified limits. The limits given for D_w apply only to the equations; for many materials, wire sizes smaller or larger than the limits shown are frequent.

Notes: (a) For *light service*, use design $s_{sd} = 0.405s_u$. For *average service*, use $s_{sd} = 0.324s_u$. For *severe service*, use $s_{sd} = 0.263s_u$. These results agree closely with Westinghouse recommendations, as reported by Wahl. (b) Equations for approximate minimum tensile strength as specified by ASTM. (c) Agrees closely with Alco recommendations; since they are higher than the stresses recommended by some authorities, a small factor of safety may be advisable, unless the spring manufacturer agrees. (d) Derived from Hunter Spring Co. data.[6.15] Value for indefinite life from 0 to max.; for 10^5 cycles for steel wire (except stainless), multiply this value by 1.4, for example. *Use minimum N* = 1.15. (e) Use both expressions for oil tempered; for *hard-drawn* wire, multiply by 0.9. (f) By analogy with music wire. Also, Associated Spring recommends the safe design range for valve spring quality as defined by the triangle ABO, Fig. 6.9, for $D_w < 0.207$ and unpeened; by triangle CBO when peened. (g) Conservative in the larger sizes. (h) In accordance with INCO.[6.13] (i) Light service, use $s_s = 0.32s_u$; average service, $0.26s_u$; severe service, $0.21s_u$. (j) Adapted from Associated Spring data. [6.2] (k) A straight-line interpolation between $s_u = 85$ ksi for $D_w = 0.5$ in. and formula limit is probably satisfactory. (l) Multiply by 0.8 for average service. (m) Stress relieved; decrease 10% if as-drawn. (n) Increase 10% if preset; 25–35% for shot peening wires larger than 0.062 in.[6.13] (o) Age hardened. (p) INCO gives 20 ksi for 10^8 cycles. (q) Multiply by 1.33 for 10^5 cycles, 0-max. (r) INCO data suggest that this K-Monel is somewhat stronger than the Monel, but detail lacking. (s) Valve spring quality.

MATERIAL	E $\times 10^{-6}$ psi	G $\times 10^{-6}$ psi	DESIGN STRESS, s_{sd} ksi. Light Load	MIN. TENSILE, s_u ksi. (Uncoiled)	MAX. "SOLID" s_s (Approx. s_{ys})	END. STR'TH. s_{no} ksi. (R = 0)
Column No. →	(1)	(2)	(3)	(4)	(5)	(6)
Oil Tempered: ASTM A229	29	11.5	(a)	$\dfrac{146}{D_w^{0.19}}$ (b) $[0.032 < D_w < 0.5]$	$0.6s_u$ (c) $[Q=87.5, x=0.19]$	$\dfrac{47}{D_w^{0.1}}$ (d)(e) $[0.041 < D_w < 0.15$
Hard Drawn: ASTM A227	29	11.5	Use 0.85 times constants in Note (a)	$\dfrac{140}{D_w^{0.19}}$ (b) $[0.028 < D_w < 0.625]$	$0.5s_u$ (c) $[Q=70, x=0.19]$	$\dfrac{30}{D_w^{0.34}}$ (d)(e) $[0.15 < D_w < 0.625$
Music Wire: ASTM A228	30	12	(a)	$\dfrac{190}{D_w^{0.154}}$ (b) $[0.004 < D_w < 0.192]$	$0.5s_u$ (c) $[Q=95, x=0.154]$ $[0.03 < D_w < 0.192;$ 190 ksi max.]	$\dfrac{50}{D_w^{0.154}}$ (d) $[0.018 < D_w < 0.1$ 92 ksi max.]

MATERIAL	$E \times 10^{-6}$ psi	$G \times 10^{-6}$ psi	DESIGN STRESS, s_{sd} ksi. Light Load	MIN. TENSILE, s_u ksi. (Uncoiled)	MAX. "SOLID" s_s (Approx. s_{ys})	END. STR'TH. s_{no} ksi. ($R=0$)
Column No. →	(1)	(2)	(3)	(4)	(5)	(6)
Carbon Steel, VSQ(s); ASTM A230	30	11.5	(a)	$\dfrac{182}{D_w^{0.1}}$ (b) $[0.093 < D_w < 0.25]$	$0.5s_u$ (c) $[Q=91, x=0.1]$ $[0.093 < D_w < 0.25]$	$\dfrac{49}{D_w^{0.15}}$ (d)(f) $[0.093 < D_w < 0.25]$
Cr-V Steel, VSQ(s); ASTM A232	30	11.5	(a)	$\dfrac{168}{D_w^{0.166}}$ (b) $[0.032 < D_w < 0.437]$	$0.6s_u$ (c) $[Q=100, x=0.166]$	Same as for A230 (g) $[0.028 < D_w < 0.5]$
Cr-Si Steel ASTM A401	29	11.5	(a)	$\dfrac{202}{D_w^{0.107}}$ (b) $[0.032 < D_w < 0.375]$	$0.6s_u$ $[Q=121, x=0.107]$	Same as for A230 (g) $[0.032 < D_w < 0.375]$
Stainless Steel (Cr-Ni) ASTM A313	26	10	(i)	$\dfrac{170}{D_w^{0.14}}$ (b) $[0.01 < D_w < 0.13]$ $\dfrac{97}{D_w^{0.41}}$ (b) $[0.13 < D_w < 0.375]$	$0.47s_u$ (h) $[Q=80, x=0.14]$ $[Q=45.6, x=0.41]$	$\dfrac{30}{D_w^{0.17}}$ (d)(q) $[0.01 < D_w < 0.375]$
Beryllium Copper	18.5	7	Use 0.8 times values for A229	160–200	$0.5s_u$ (h)	$\dfrac{35}{D_w^{0.2}}$ (d) $[0.09 < D_w < 0.5;$ 56 ksi max.]
Spring Brass	14.5	4.5	Use 0.35 times values for A229	$\dfrac{88}{D_w^{0.1}}$ (j)(k) $[0.03 < D_w < 0.20;$ 125 ksi max.]	$\dfrac{42}{D_w^{0.26}}$ (j) $[0.08 < D_w < 0.5;$ 68 ksi max.]	$\dfrac{11.5}{D_w^{0.2}}$ (d) $[0.09 < D_w < 0.5;$ 19 ksi max.]
Phosphor Bronze	14.5	6	Use 0.5 times values for A229	$\dfrac{106}{D_w^{0.08}}$ (j) $[D_w \le 0.5;$ 145 ksi max.]	$0.45s_u$ (h) $[Q=47.5; x=0.08]$	$\dfrac{15.3}{D_w^{0.2}}$ (d) $[0.09 < D_w < 0.5;$ 28 ksi max.]
Monel (m)	24.5	9.3	$\dfrac{52}{D_w^{0.1}}$ (h)(l) $[0.058 < D_w < 0.625;$ 70 ksi max.]	$\dfrac{129}{D_w^{0.1}}$ (h) $[D_w \le 0.625;$ 170 ksi max.]	$0.4s_u$ (h)(n) $[Q=51.5; x=0.1]$ $[D_w \le 0.625;$ 68 ksi max.]	$\dfrac{18}{D_w^{0.2}}$ (d)(n) $[D_w \le 0.625;$ 29 ksi max.(p)]
K-Monel (o)	24.5	9.3	75 ksi (h)(l) 80 ksi, preset $[D_w > 0.058]$	$\dfrac{158}{D_w^{0.048}}$ (h) $[D_w \le 0.625;$ 180 ksi max.]	$0.4s_u$ (h)(n) $[Q=63, x=0.048]$ $[D_w \le 0.625;$ 72 ksi max.]	Use same as Monel (r) [29 ksi max. (p)]

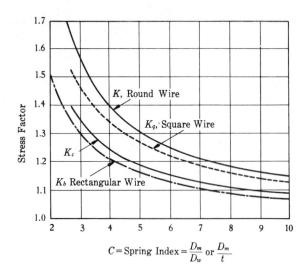

$$C = \text{Spring Index} = \frac{D_m}{D_w} \text{ or } \frac{D_m}{t}$$

FIGURE AF 15 Stress Factors (Wahl). Use D_m/D_w as the spring index for round-wire springs, D_m/t for rectangular-wire springs, where t is the dimension perpendicular to the axis of the spring. It is advisable that D_m/D_w be not less than 3, preferably greater than 4. (After R. E. Peterson).[4.21]

TABLE AT 18 STRESS FACTORS FOR CURVED BEAMS

Extracted with permission of publisher, John Wiley & Sons, Inc., from Seely and Smith.[1·7] For a hollow circular section, the values for a solid section may be used with little error when $r/c \geqq 1.8$; K_{ci} for point on inside of curvature: K_{co} for outside.

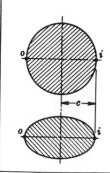

← Center of Curvature

$r/c \rightarrow$	1.2	1.4	1.6	1.8	2.0	3.0	4.0	6.0	8.0	10.0
K_{ci}	3.41	2.40	1.96	1.75	1.62	1.33	1.23	1.14	1.10	1.08
K_{co}	0.54	0.60	0.65	0.68	0.71	0.79	0.84	0.89	0.91	0.93
y_o/r	0.22	0.15	0.11	0.08	0.07	0.03	0.02	0.01		

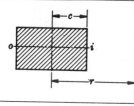

K_{ci}	2.89	2.13	1.79	1.63	1.52	1.3	1.2	1.12	1.09	1.07
K_{co}	0.57	0.63	0.67	0.70	0.73	0.81	0.85	0.9	0.92	0.94
y_o/r	0.31	0.20	0.15	0.11	0.09	0.04	0.02	0.01		

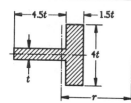

K_{ci}	3.63	2.54	2.14	1.89	1.73	1.41	1.29	1.18	1.13	1.10
K_{co}	0.58	0.63	0.67	0.70	0.72	0.79	0.83	0.88	0.91	0.92
y_o/r	0.42	0.30	0.23	0.18	0.15	0.07	0.04	0.02	0.01	

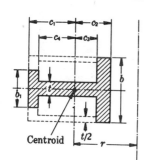

Centroid

$$Z = -1 + \frac{r}{A}[b_1 \log_e(r + c_1) + (t - b_1)\log_e(r + c_4)$$
$$+ (b - t)\log_e(r - c_3) - b \log_e(r - c_2)].$$

For a **T-section** let $c_4 = c_1$ and $b_1 \doteq t$.

For an **I-section**, let $b_1 = b$. Also, if the front and rear flanges are the same in thickness, c_2 will be equal to c_1 and c_3 will be equal to c_4.

For a **box-section** (dotted outline), Z is the same as for an I-section, each side panel of the box being $t/2$ inches thick.

Flanges in compression should not be so thin as to result in local buckling.

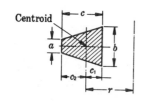

Centroid

$$Z = -1 + \frac{2r}{(a + b)c}\left\{\left[a + \frac{b - a}{c}(r + c_2)\right]\right.$$
$$\left. \times \log_e\left(\frac{r + c_2}{r - c_1}\right) - (b - a)\right\}$$

TABLE AT 19   KEY DIMENSIONS

See Figs. 10.1 and 10.2 for b and t. From ASA standard B17.1-1943. Other sizes available. The tolerances on t may be numerically the same as given, negative on plain keys, positive on tapered keys.

SHAFT DIAMETER (*Inclusive*)	b	t	TOLERANCE ON b, *in.*
$\frac{1}{2} - \frac{9}{16}$	$\frac{1}{8}$	$\frac{3}{32}$	-0.0020
$\frac{5}{8} - \frac{7}{8}$	$\frac{3}{16}$	$\frac{1}{8}$	-0.0020
$\frac{15}{16}-1\frac{1}{4}$	$\frac{1}{4}$	$\frac{3}{16}$	-0.0020
$1\frac{5}{16}-1\frac{3}{8}$	$\frac{5}{16}$	$\frac{1}{4}$	-0.0020
$1\frac{7}{16}-1\frac{3}{4}$	$\frac{3}{8}$	$\frac{1}{4}$	-0.0020
$1\frac{13}{16}-2\frac{1}{4}$	$\frac{1}{2}$	$\frac{3}{8}$	-0.0025
$2\frac{5}{16}-2\frac{3}{4}$	$\frac{5}{8}$	$\frac{7}{16}$	-0.0025
$2\frac{7}{8} -3\frac{1}{4}$	$\frac{3}{4}$	$\frac{1}{2}$	-0.0025
$3\frac{3}{8} -3\frac{3}{4}$	$\frac{7}{8}$	$\frac{5}{8}$	-0.0030
$3\frac{7}{8} -4\frac{1}{2}$	1	$\frac{3}{4}$	-0.0030
$4\frac{3}{4} -5\frac{1}{2}$	$1\frac{1}{4}$	$\frac{7}{8}$	-0.0030
$5\frac{3}{4} -6$	$1\frac{1}{2}$	1	-0.0030

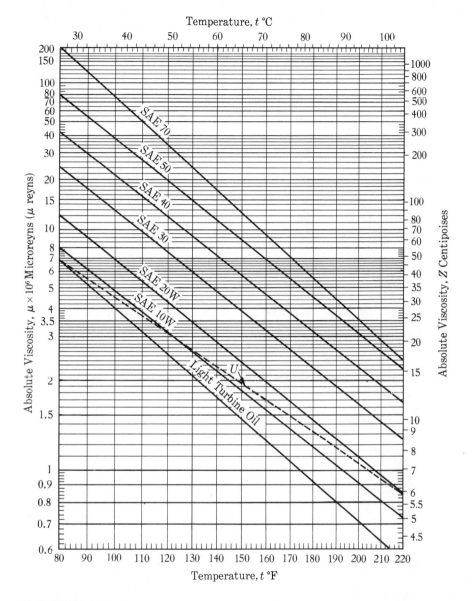

FIGURE AF 16 Typical Viscosities of Oils. For ring-oiled bearings, usually SAE 20 (or the equivalent) or lighter. SAE 70 and chart paper by courtesy of Westinghouse Electric Corp. Dotted curve U is for a high viscosity-index oil, Uniflow—typical test values, Standard Oil of N.J. Other data from The Texas Co. On average, an SAE 10W–30 oil has a viscosity a little lower than SAE 30 at 210°F, a little higher than SAE 10 at 100°F.

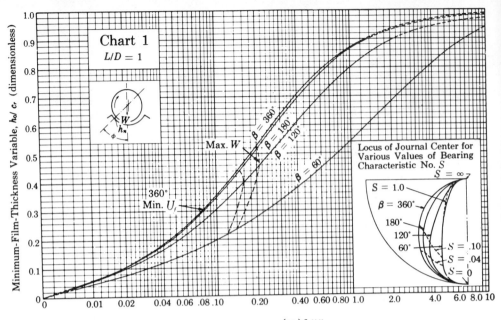

FIGURE AF 17 Minimum-Film Variable vs. Sommerfeld S (with Side Flow). (Courtesy Raimondi and Boyd[11.7] and Westinghouse Electric).

FIGURE AF 18 Coefficient-of-Friction Variable vs. Sommerfeld S (with Side Flow). (Courtesy Raimondi and Boyd[11.7] and Westinghouse Electric).

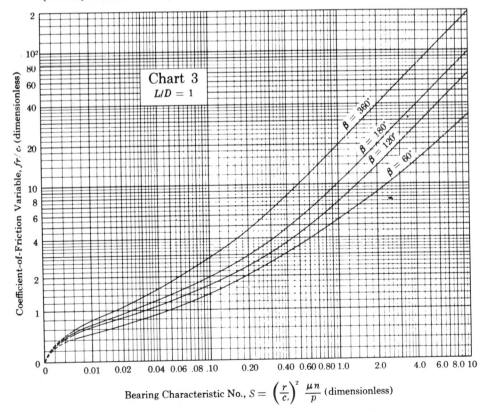

TABLE AT 20 DIMENSIONLESS PERFORMANCE PARAMETERS FOR FULL JOURNAL BEARINGS WITH SIDE FLOW

Courtesy Raimondi and Boyd[11.7] and Westinghouse Electric. Values of h_o/c_r for optimum bearings, maximum load and minimum friction, respectively: for $L/D = \infty$, 0.66, 0.60; for $L/D = 1$, 0.53, 0.30; for $L/D = 0.5$, 0.43, 0.12; for $L/D = 0.25$, 0.27, 0.03.

L/D	ϵ	$\dfrac{h_o}{c_r}$	S	ϕ	$\dfrac{r}{c_r}f$	$\dfrac{q}{rc_r n_s L}$	$\dfrac{q_s}{q}$	$\dfrac{\rho c \Delta t_o}{p}$	$\dfrac{p}{p_{\max}}$
∞	0	1.0	∞	(70.92)	∞	π	0	∞	—
	0.1	0.9	0.240	69.10	4.80	3.03	0	19.9	0.826
	0.2	0.8	0.123	67.26	2.57	2.83	0	11.4	0.814
	0.4	0.6	0.0626	61.94	1.52	2.26	0	8.47	0.764
	0.6	0.4	0.0389	54.31	1.20	1.56	0	9.73	0.667
	0.8	0.2	0.021	42.22	0.961	0.760	0	15.9	0.495
	0.9	0.1	0.0115	31.62	0.756	0.411	0	23.1	0.358
	0.97	0.03	—	—	—	—	0	—	—
	1.0	0	0	0	0	0	0	∞	0
1	0	1.0	∞	(85)	∞	π	0	∞	—
	0.1	0.9	1.33	79.5	26.4	3.37	0.150	106	0.540
	0.2	0.8	0.631	74.02	12.8	3.59	0.280	52.1	0.529
	0.4	0.6	0.264	63.10	5.79	3.99	0.497	24.3	0.484
	0.6	0.4	0.121	50.58	3.22	4.33	0.680	14.2	0.415
	0.8	0.2	0.0446	36.24	1.70	4.62	0.842	8.00	0.313
	0.9	0.1	0.0188	26.45	1.05	4.74	0.919	5.16	0.247
	0.97	0.03	0.00474	15.47	0.514	4.82	0.973	2.61	0.152
	1.0	0	0	0	0	—	1.0	0	0
$\frac{1}{2}$	0	1.0	∞	(88.5)	∞	π	0	∞	—
	0.1	0.9	4.31	81.62	85.6	3.43	0.173	343.0	0.523
	0.2	0.8	2.03	74.94	40.9	3.72	0.318	164.0	0.506
	0.4	0.6	0.779	61.45	17.0	4.29	0.552	68.6	0.441
	0.6	0.4	0.319	48.14	8.10	4.85	0.730	33.0	0.365
	0.8	0.2	0.0923	33.31	3.26	5.41	0.874	13.4	0.267
	0.9	0.1	0.0313	23.66	1.60	5.69	0.939	6.66	0.206
	0.97	0.03	0.00609	13.75	0.610	5.88	0.980	2.56	0.126
	1.0	0	0	0	0	—	1.0	0	0
$\frac{1}{4}$	0.0	1.0	∞	(89.5)	∞	π	0	∞	—
	0.1	0.9	16.2	82.31	322.0	3.45	0.180	1287.0	0.515
	0.2	0.8	7.57	75.18	153.0	3.76	0.330	611.0	0.489
	0.4	0.6	2.83	60.86	61.1	4.37	0.567	245.0	0.415
	0.6	0.4	1.07	46.72	26.7	4.99	0.746	107.0	0.334
	0.8	0.2	0.261	31.04	8.80	5.60	0.884	35.4	0.240
	0.9	0.1	0.0736	21.85	3.50	5.91	0.945	14.1	0.180
	0.97	0.03	0.0101	12.22	0.922	6.12	0.984	3.73	0.108
	1.0	0	0	0	0	—	1.0	0	0

q in.3/sec.　　　$\rho \approx 0.03$ lb/in.^{3}　　　$c = 3734$ in–lb/lb–°F.　　　$\rho c = 112$.

TABLE AT 21 DIMENSIONLESS PERFORMANCE PARAMETERS FOR 180° BEARING, CENTRALLY LOADED, WITH SIDE FLOW

Courtesy Raimondi and Boyd[11.7] and Westinghouse Electric. Values of h_o/c_r for optimum bearings, maximum load and minimum friction, respectively: for $L/D = \infty$, 0.64, 0.6; for $L/D = 1$, 0.52, 0.44; for $L/D = 0.5$, 0.42, 0.23; for $L/D = 0.25$, 0.28, 0.03.

L/D	ϵ	$\dfrac{h_o}{c_r}$	S	ϕ	$\dfrac{r}{c_r}f$	$\dfrac{q}{rc_r n_s L}$	$\dfrac{q_s}{q}$	$\dfrac{\rho c \Delta t_o}{p}$	$\dfrac{p}{p_{max}}$
∞	0	1.0	∞	90.0	∞	π	∞	∞	—
	0.1	0.9	0.347	72.90	3.55	3.04	0	14.7	0.778
	0.2	0.8	0.179	61.32	2.01	2.80	0	8.99	0.759
	0.4	0.6	0.0898	49.99	1.29	2.20	0	7.34	0.700
	0.6	0.4	0.0523	43.15	1.06	1.52	0	8.71	0.607
	0.8	0.2	0.0253	33.35	0.859	0.767	0	14.1	0.459
	0.9	0.1	0.0128	25.57	0.681	0.380	0	22.5	0.337
	0.97	0.03	0.00384	15.43	0.416	0.119	0	44.0	0.190
	1.0	0	0	0	0	0	0	∞	0
1	0	1.0	∞	90.0	—	π	0	∞	—
	0.1	0.9	1.40	78.50	14.1	3.34	0.139	57.0	0.525
	0.2	0.8	0.670	68.93	7.15	3.46	0.252	29.7	0.513
	0.4	0.6	0.278	58.86	3.61	3.49	0.425	16.5	0.466
	0.6	0.4	0.128	44.67	2.28	3.25	0.572	12.4	0.403
	0.8	0.2	0.0463	32.33	1.39	2.63	0.721	10.4	0.313
	0.9	0.1	0.0193	24.14	0.921	2.14	0.818	9.13	0.244
	0.97	0.03	0.00483	14.57	0.483	1.60	0.915	6.96	0.157
	1.0	0	0	0	0	—	1.0	0	0
$\frac{1}{2}$	0	1.0	∞	90.0	∞	π	0	∞	—
	0.1	0.9	4.38	79.97	44.0	3.41	0.167	177.0	0.518
	0.2	0.8	2.06	72.14	21.6	3.64	0.302	87.8	0.499
	0.4	0.6	0.794	58.01	9.96	3.93	0.506	42.7	0.438
	0.6	0.4	0.321	45.01	5.41	3.93	0.665	25.9	0.365
	0.8	0.2	0.0921	31.29	2.54	3.56	0.806	15.0	0.273
	0.9	0.1	0.0314	22.80	1.38	3.17	0.886	9.80	0.208
	0.97	0.03	0.00625	13.63	0.581	2.62	0.951	5.30	0.132
	1.0	0	0	0	0	—	1.0	0	0
$\frac{1}{4}$	0	1.0	∞	90.0	∞	π	0	∞	—
	0.1	0.9	16.3	81.40	163.0	3.44	0.176	653.0	0.513
	0.2	0.8	7.60	73.70	79.4	3.71	0.320	320.0	0.489
	0.4	0.6	2.84	58.99	35.1	4.11	0.534	146.0	0.417
	0.6	0.4	1.08	44.96	17.6	4.25	0.698	79.8	0.336
	0.8	0.2	0.263	30.43	6.88	4.07	0.837	36.5	0.241
	0.9	0.1	0.0736	21.43	2.99	3.72	0.905	18.4	0.180
	0.97	0.03	0.0104	12.28	0.877	3.29	0.961	6.46	0.110
	1.0	0	0	0	0	—	1.0	0	0

TABLE AT 22 DIMENSIONLESS PERFORMANCE PARAMETERS FOR 120° BEARING, CENTRALLY LOADED, WITH SIDE FLOW.

Courtesy Raimondi and Boyd[11.7] and Westinghouse Electric. Values of h_o/c_r for optimum bearings, maximum load and minimum friction, respectively: for $L/D = \infty$, 0.53, 0.5; for $L/D = 1$, 0.46, 0.4; for $L/D = 0.5$, 0.38, 0.28; for $L/D = 0.25$, 0.26, 0.06. When $1 - h_o/c_r \neq \epsilon$, the trailing end of the bearing does not reach h_o as defined in Fig. 11.6; that is, h_o in this table is h_{min}.

L/D	ϵ	$\dfrac{h_o}{c_r}$	S	ϕ	$\dfrac{r}{c_r}f$	$\dfrac{q}{rc_r n_s L}$	$\dfrac{q_s}{q}$	$\dfrac{\rho c \Delta t_o}{p}$	$\dfrac{p}{p_{max}}$
∞	0	1.0	∞	90.0	∞	π	0	∞	—
	0.1	0.9007	0.877	66.69	6.02	3.02	0	25.1	0.610
	0.2	0.8	0.431	52.60	3.26	2.75	0	14.9	0.599
	0.4	0.6	0.181	39.02	1.78	2.13	0	10.5	0.566
	0.6	0.4	0.0845	32.67	1.21	1.47	0	10.3	0.509
	0.8	0.2	0.0328	26.80	0.853	0.759	0	14.1	0.405
	0.9	0.1	0.0147	21.51	0.653	0.388	0	21.2	0.311
	0.97	0.03	0.00406	13.86	0.399	0.118	0	42.4	0.199
	1.0	0	0	0	0	0	0	∞	0
1	0	1.0	∞	90.0	∞	π	0	∞	—
	0.1	0.9024	2.14	72.43	14.5	3.20	0.0876	59.5	0.427
	0.2	0.8	1.01	58.25	7.44	3.11	0.157	32.6	0.420
	0.4	0.6	0.385	43.98	3.60	2.75	0.272	19.0	0.396
	0.6	0.4	0.162	35.65	2.16	2.24	0.384	15.0	0.356
	0.8	0.2	0.0531	27.42	1.27	1.57	0.535	13.9	0.290
	0.9	0.1	0.0208	21.29	0.855	1.11	0.657	14.4	0.233
	0.97	0.03	0.00498	13.49	0.461	0.694	0.812	14.0	0.162
	1.0	0	0	0	0	—	1.0	0	0
$\frac{1}{2}$	0	1.0	∞	90.0	∞	π	0	—	—
	0.1	0.9034	5.42	74.99	36.6	3.29	0.124	149.0	0.431
	0.2	0.8003	2.51	63.38	18.1	3.32	0.225	77.2	0.424
	0.4	0.6	0.914	48.07	8.20	3.15	0.386	40.5	0.389
	0.6	0.4	0.354	38.50	4.43	2.80	0.530	27.0	0.336
	0.8	0.2	0.0973	28.02	2.17	2.18	0.684	19.0	0.261
	0.9	0.1	0.0324	21.02	1.24	1.70	0.787	15.1	0.203
	0.97	0.03	0.00631	13.00	0.550	1.19	0.899	10.6	0.136
	1.0	0	0	0	0	—	1.0	0	0
$\frac{1}{4}$	0	1.0	∞	90.0	∞	π	0	∞	—
	0.1	0.9044	18.4	76.97	124.0	3.34	0.143	502.0	0.456
	0.2	0.8011	8.45	65.97	60.4	3.44	0.260	254.0	0.438
	0.4	0.6	3.04	51.23	26.6	3.42	0.442	125.0	0.389
	0.6	0.4	1.12	40.42	13.5	3.20	0.599	75.8	0.321
	0.8	0.2	0.268	28.38	5.65	2.67	0.753	42.7	0.237
	0.9	0.1	0.0743	20.55	2.63	2.21	0.846	25.9	0.178
	0.97	0.03	0.0105	12.11	0.832	1.69	0.931	11.6	0.112
	1.0	0	0	0	0	—	1.0	0	0

TABLE AT 23 DIMENSIONLESS PERFORMANCE PARAMETERS FOR 60° BEARING, CENTRALLY LOADED, WITH SIDE FLOW

Courtesy Raimondi and Boyd[11.7] and Westinghouse Electric. Values of h_o/c_r for optimum bearings, maximum load and minimum friction, respectively: for $L/D = \infty$, 0.25, 0.23; for $L/D = 1$, 0.23, 0.22; for $L/D = 0.5$, 0.2, 0.16; for $L/D = 0.25$, 0.15, 0.1. When $1 - h_o/c_r \neq \epsilon$, the trailing end of the bearing does not reach h_o as defined in Fig. 11.6; that is, h_o in this table is $h_{\min}$.

L/D	ϵ	$\dfrac{h_o}{c_r}$	S	ϕ	$\dfrac{r}{c_r}f$	$\dfrac{q}{rc_r n_s L}$	$\dfrac{q}{q_s}$	$\dfrac{\rho c \Delta t_o}{p}$	$\dfrac{p}{p_{\max}}$
∞	0	1.0	∞	90.0	∞	π	0	∞	—
	0.1	0.9191	5.75	65.91	19.7	3.01	0	82.3	0.337
	0.2	0.8109	2.66	48.91	10.1	2.73	0	46.5	0.336
	0.4	0.6002	0.931	31.96	4.67	2.07	0	28.4	0.329
	0.6	0.4	0.322	23.21	2.40	1.40	0	21.5	0.317
	0.8	0.2	0.0755	17.39	1.10	0.722	0	19.2	0.287
	0.9	0.1	0.0241	14.94	0.667	0.372	0	22.5	0.243
	0.97	0.03	0.00495	10.58	0.372	0.115	0	40.7	0.163
	1.0	0	0	0	0	0	0	∞	0
1	0	1.0	∞	90.0	∞	π	0	∞	—
	0.1	0.9212	8.52	67.92	29.1	3.07	0.0267	121.0	0.252
	0.2	0.8133	3.92	50.96	14.8	2.82	0.0481	67.4	0.251
	0.4	0.6010	1.34	33.99	6.61	2.22	0.0849	39.1	0.247
	0.6	0.4	0.450	24.56	3.29	1.56	0.127	28.2	0.239
	0.8	0.2	0.101	18.33	1.42	0.883	0.200	22.5	0.220
	0.9	0.1	0.0309	15.33	0.822	0.519	0.287	23.2	0.192
	0.97	0.03	0.00584	10.88	0.422	0.226	0.465	30.5	0.139
	1.0	0	0	0	0	—	1.0	0	0
$\frac{1}{2}$	0	1.0	∞	90.0	∞	π	0.0	∞	—
	0.1	0.9223	14.2	69.00	48.6	3.11	0.0488	201.0	0.239
	0.2	0.8152	6.47	52.60	24.2	2.91	0.0883	109.0	0.239
	0.4	0.6039	2.14	37.00	10.3	2.38	0.160	59.4	0.233
	0.6	0.4	0.695	26.98	4.93	1.74	0.236	40.3	0.225
	0.8	0.2	0.149	19.57	2.02	1.05	0.350	29.4	0.201
	0.9	0.1	0.0422	15.91	1.08	0.664	0.464	26.5	0.172
	0.97	0.03	0.00704	10.85	0.490	0.329	0.650	27.8	0.122
	1.0	0	0	0	0	—	1.0	0	0
$\frac{1}{4}$	0	1.0	∞	90.0	∞	π	0	∞	—
	0.1	0.9251	35.8	71.55	121.0	3.16	0.0666	499.0	0.251
	0.2	0.8242	16.0	58.51	58.7	3.04	0.131	260.0	0.249
	0.4	0.6074	5.20	41.01	24.5	2.57	0.236	136.0	0.242
	0.6	0.4	1.65	30.14	11.2	1.98	0.346	86.1	0.228
	0.8	0.2	0.333	21.70	4.27	1.30	0.496	54.9	0.195
	0.9	0.1	0.0844	16.87	2.01	0.894	0.620	41.0	0.159
	0.97	0.03	0.0110	10.81	0.713	0.507	0.786	29.1	0.107
	1.0	0	0	0	0	—	1.0	0	0

TABLE AT 24   VALUES OF FORM FACTOR Y IN LEWIS' EQUATION

FD = full depth.

NO. TEETH	LOAD AT TIP $14\frac{1}{2}°$ FD	LOAD AT TIP 20° FD	LOAD AT TIP 20° Stub	LOAD NEAR MIDDLE $14\frac{1}{2}°$ FD	LOAD NEAR MIDDLE 20° FD	NO. TEETH	LOAD AT TIP $14\frac{1}{2}°$ FD	LOAD AT TIP 20° FD	LOAD AT TIP 20° Stub	LOAD NEAR MIDDLE $14\frac{1}{2}°$ FD	LOAD NEAR MIDDLE 20° FD
10	0.176	0.201	0.261			32	0.322	0.364	0.443	0.547	0.617
11	0.192	0.226	0.289			33	0.324	0.367	0.445	0.550	0.623
12	0.210	0.245	0.311	0.355	0.415	35	0.327	0.373	0.449	0.556	0.633
13	0.223	0.264	0.324	0.377	0.443	37	0.330	0.380	0.454	0.563	0.645
14	0.236	0.276	0.339	0.399	0.468	39	0.335	0.386	0.457	0.568	0.655
15	0.245	0.289	0.349	0.415	0.490	40	0.336	0.389	0.459	0.570	0.659
16	0.255	0.295	0.360	0.430	0.503	45	0.340	0.399	0.468	0.579	0.678
17	0.264	0.302	0.368	0.446	0.512	50	0.346	0.408	0.474	0.588	0.694
18	0.270	0.308	0.377	0.459	0.522	55	0.352	0.415	0.480	0.596	0.704
19	0.277	0.314	0.386	0.471	0.534	60	0.355	0.421	0.484	0.603	0.713
20	0.283	0.320	0.393	0.481	0.544	65	0.358	0.425	0.488	0.607	0.721
21	0.289	0.326	0.399	0.490	0.553	70	0.360	0.429	0.493	0.610	0.728
22	0.292	0.330	0.404	0.496	0.559	75	0.361	0.433	0.496	0.613	0.735
23	0.296	0.333	0.408	0.502	0.565	80	0.363	0.436	0.499	0.615	0.739
24	0.302	0.337	0.411	0.509	0.572	90	0.366	0.442	0.503	0.619	0.747
25	0.305	0.340	0.416	0.515	0.580	100	0.368	0.446	0.506	0.622	0.755
26	0.308	0.344	0.421	0.522	0.588	150	0.375	0.458	0.518	0.635	0.779
27	0.311	0.348	0.426	0.528	0.592	200	0.378	0.463	0.524	0.640	0.787
28	0.314	0.352	0.430	0.534	0.597	300	0.382	0.471	0.534	0.650	0.801
29	0.316	0.355	0.434	0.537	0.602	Rack	0.390	0.484	0.550	0.660	0.823
30	0.318	0.358	0.437	0.540	0.606						

"STANDARD" DIAMETRAL PITCHES

2,   2.25,   2.5,   3,   3.5,   4,   5,   6,   7,   8,   10,   12,   16

TABLE AT 25 VALUES OF C FOR e = 0.001 in.

For other values of e, multiply the value given by the number of thousandths that e is; for example, for cast iron and cast iron, $14\frac{1}{2}°$ full depth, and $e = 0.004$ in., $C = (4)(800) = 3200$. For other materials, use $C = kE_gE_p/(E_g + E_p)$. Values of C for bronze are virtually the same as for the gray iron (modulii of elasticity about the same). FD = full depth.

MATERIAL	$14\frac{1}{2}°$ FD $(k = 0.107e)$	20° FD $(k = 0.111e)$	20° STUB $(k = 0.115e)$
Gray iron and gray iron . .	800	830	860
Gray iron and steel . . .	1100	1140	1180
Steel and steel	1600	1660	1720

143

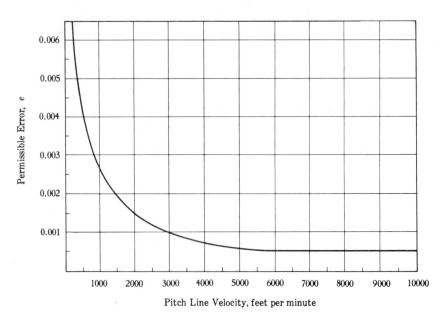

FIGURE AF 19 Maximum Permissible Errors in Gear-tooth Profiles. Extreme quietness will require a smaller error than indicated by this curve.

FIGURE AF 20 Expected Errors in Tooth Profiles. (Data courtesy Earle Buckingham).

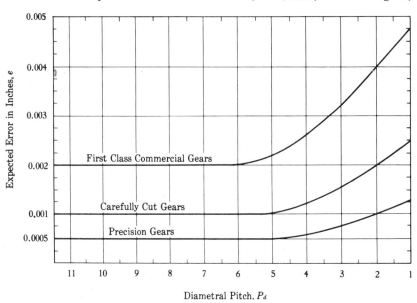

TABLE AT 26   VALUES OF LIMITING WEAR-LOAD FACTOR K_g

Specified BHN's are minimums. Values are for indefinite life unless otherwise indicated. Straight line interpolations on the sum of BHN's are permissible when the difference in BHN's is less than 100 points.

COMBINATIONS OF MATERIALS (BHN) AND LIFE	s_{nsurf} ksi	K_g $14\frac{1}{2}°$	$20°$
BOTH GEARS STEEL:			
Sum of BHN = 300, 10^6 cycles		63	86
Ditto 10^7 cycles		40	54
Ditto 4×10^7 cycles or more	50	30	41
Sum of BHN = 350	60	43	58
Sum of BHN = 400, 10^6 cycles		119	162
Ditto 10^7 cycles		75	102
Ditto 4×10^7 cycles or more	70	58	79
Sum of BHN = 450	80	76	103
Sum of BHN = 500	90	96	131
Sum of BHN = 550	100	119	162
Sum of BHN = 600, 10^6 cycles		292	400
Ditto 10^7 cycles		185	252
Ditto 4×10^7 cycles or more	110	144	196
Sum of BHN = 650	120	171	233
Sum of BHN = 700	130	196	270
Sum of BHN = 750	140	233	318
Sum of BHN = 800	150	268	366
Steel (500) and steel (350)	145	250	342
Steel (450) and same	170	344	470
Steel (500), induction hardened, and same, 10^7 cycles . .		880	1190
Ditto 10^8 cycles . .		670	920
Ditto 10^{10} cycles . .		405	555
Steel (600), carburized case hardened, and same, 10^7 cycles .		1230	1680
Ditto 10^8 cycles .		940	1280
Ditto 10^{10} cycles .		550	750
Steel (150) and cast iron	50	44	60
Steel (250) and Ni cast iron, HT	90	150	205
Steel (630) and SAE 65 phosphor bronze (67)*		53	72
Steel (250 and over) and chilled phosphor bronze	83	128	175
Steel (630) and laminated phenolic*		46	64
Cast iron, class 20, and same*		81	112
Cast iron and same, 10^6 cycles		376	515
Ditto 10^7 cycles		212	290
Ditto 4×10^7 cycles†		150	205
Cast iron with steel scrap and same		170	230
Cast iron, class 30, austempered (270) and same* . . .		224	306
G. M. Meehanite (190) and same*		104	142
Nodular iron casting, 80–60–03 (210) and same* . . .		180	248
Cast iron and phosphor bronze.	83	170	234
Cast iron, class 30 (340) and cast aluminum, SAE 39 (60)* . .		16	22

* These values adapted from Cram.[12.22]

† These values are not consistent with those from Cram, and probably should be discounted.

TABLE AT 27   WEAR FACTOR K_w FOR WORM GEARS

Taken from Buckingham and Ryffel,[14.1] with permission of the publisher, The Industrial Press.

MATERIALS		THREAD ANGLE ϕ_n			
Worm	*Gear*	$14\frac{1}{2}°$	20°	25°	30°
†Hardened steel	Chilled bronze	90	125	150	180
†Hardened steel	Bronze	60	80	100	120
Steel, 250 BHN (min.)	Bronze	36	50	60	72
High-test C.I.	Bronze	80	115	140	165
*Gray iron	Aluminum	10	12	15	18
*High-test C.I.	Gray iron	90	125	150	180
High-test C.I.	Cast steel	22	31	37	45
High-test C.I.	High-test C.I.	135	185	225	270
Steel 250 BHN (min.)	Laminated phenolic	47	64	80	95
Gray iron	Laminated phenolic	70	96	120	140

* For steel worms, multiply given values by 0.6.   † Over 500 BHN surface.

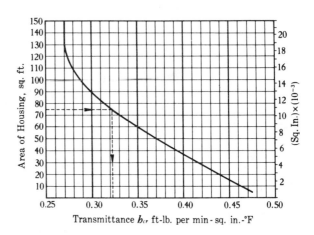

Transmittance h_{cr} ft-lb. per min-sq. in.-°F

FIGURE AF 21 Transmittance, Gear Cases. To use: Determine the area of the housing; enter chart with this area, move rightward to curve, downward to abscissa, which is the transmittance h_{cr}—as shown by dotted lines.

TABLE AT 28 PROPERTIES OF WIRE ROPE

Approximate weight of rope is w lb. per ft.; D_r = diameter of rope; D_s = diameter of sheave; A_m = cross-section area of metal, sq. in.; E_r = modulus of elasticity of the rope; I.P.S. stands for improved plow steel; P.S. for plow steel; M.P.S. for mild plow steel. All values are for rope with fiber core. It may occasionally be useful in *preliminary* computations to *estimate* the ultimate strengths of terms in D_r; for example, for 6×19 rope: VHS, $F_u \approx 48D_r^2$ tons; IPS, $F_u \approx 42D_r^2$; PS $F_u \approx 36D_r^2$; MPS $F_u \approx 32D_r^2$ tons. Multiply the values given by 1.075 to obtain strength of IWRC. For 6×19 traction steel, $F_u = (0.87)(F_u$ for MPS) will be safe.

DIA. OF ROPE D_r., in.	6×7 Wire Rope $w = 1.52D_r^2$ lb/ft. Min. $D_s = 42D_r$ in. Desirable $D_s = 72D_r$ in. $D_w \approx 0.111D_r$ $A_m \approx 0.38D_r^2$ sq. in. $E_r \approx 13 \times 10^6$ psi			6×19 Wire Rope $w \approx 1.6D_r^2$ lb/ft. Min. $D_s = 30D_r$ in. Desirable $D_s = 45D_r$ in. $D_w \approx 0.067D_r$ $A_m \approx 0.4D_r^2$ sq. in. $E_r \approx 12 \times 10^6$ psi IWRC: $w \approx 1.76D_r^2$ lb/ft.			6×37 Wire Rope $w = 1.55D_r^2$ Min. $D_s = 18D_r$ in. Desirable $D_s = 27D_r$ in. $D_w \approx 0.048D_r$ in. $A_m \approx 0.4D_r^2$ $E_r \approx 12 \times 10^6$ IWRC: $w \approx 1.71D_r^2$	
	NOMINAL BREAKING STRENGTH IN TONS OF 2000 LB., F_u							
	I.P.S.	P.S.	M.P.S.	I.P.S.	P.S.	M.P.S.	I.P.S.	P.S.
$\frac{1}{4}$	2.64	2.30	2.00	2.74	2.39	2.07	2.59	2.25
$\frac{5}{16}$	4.10	3.56	3.10	4.26	3.71	3.22	4.03	3.50
$\frac{3}{8}$	5.86	5.10	4.43	6.10	5.31	4.62	5.77	5.02
$\frac{7}{16}$	7.93	6.90	6.00	8.27	7.19	6.25	7.82	6.80
$\frac{1}{2}$	10.3	8.96	7.79	10.7	9.35	8.13	10.2	8.85
$\frac{9}{16}$	13.0	11.3	9.82	13.5	11.8	10.2	12.9	11.2
$\frac{5}{8}$	15.9	13.9	12.0	16.7	14.5	12.6	15.8	13.7
$\frac{3}{4}$	22.7	19.8	17.2	23.8	20.7	18.0	22.6	19.6
$\frac{7}{8}$	30.7	26.7	23.2	32.2	28.0	24.3	30.6	26.6
1	39.7	34.5	30.0	41.8	36.4	31.6	39.8	34.6
$1\frac{1}{8}$	49.8	43.3	37.7	52.6	45.7	39.8	50.1	43.5
$1\frac{1}{4}$	61.0	53.0	46.1	64.6	56.2	48.8	61.5	53.5
$1\frac{3}{8}$	73.1	63.6	55.3	77.7	67.5	58.8	74.1	64.5
$1\frac{1}{2}$	86.2	75.0	65.2	92.0	80.0	69.6	87.9	76.4
$1\frac{5}{8}$				107.0	93.4	81.2	103.0	89.3
$1\frac{3}{4}$				124.0	108.0	93.6	119.0	103.0
$1\frac{7}{8}$				141.0	123.0	107.0	136.0	118.0
2				160.0	139.0	121.0	154.0	134.0
$2\frac{1}{8}$				179.0	156.0		173.0	150.0
$2\frac{1}{4}$				200.0	174.0		193.0	168.0
$2\frac{1}{2}$				244.0	212.0		236.0	205.0
$2\frac{3}{4}$				292.0	254.0		284.0	247.0

From mixed references, mostly *(18.8, 18.11, 18.13, 18.22)*. Unless otherwise indicated, values of *f* are for dry surfaces; if greasy or wet, they are much lower. If *f* at the lower end of the range (or below) is used in design, this in effect introduces a design factor. Use the lower pressure *P* given as the design maximum pressure where possible. The drum temperatures are the maximum values for steady operation. See manufacturers' catalogs for more detail.

MATERIAL	DRUM MAX. t, °F	f	P PSI (*max.*)	MAX. VEL. *fpm*
Metal on metal		0.2 to 0.25	150 (250)	
Wood on metal	150	0.2 to 0.25	50 (90)	
Leather on metal or wood	150	0.3 to 0.4	15 (40)	
Cork on iron		0.35	10 (15)	
Molded blocks	650		150	7500
Asbestos in rubber compound, compressed,				
on metal	400	0.3 to 0.4	75 (100)	
Asbestos in resin binder, molded, on metal.	500	0.3 to 0.4	75 (100)	5000
in oil		≈0.10	(600)	
Asbestos, flexible woven, on metal . .	300	0.35 to 0.45	50	
in oil		≈0.12		
Sintered metal on cast iron	> 400	0.20 to 0.40	400	
in oil (as in automatic transmissions) .		0.05 to 0.08		

KIND OF WELD AND STRESS	CODES, *ksi*
BUTTWELDS	
Tension	20(b)
	8(t)
Compression . . .	20(b)
	18(c)
Shear	13(b)
Bending.	
FILLET WELDS	
All	13.6(h)
Parallel	13.6(b)
	12.4(c)
Transverse	

Reinforced butt weld, $K_f = 1.2$
T-weld, sharp corners, $K_f = 2.0$

Toe of transverse fillet weld, $K_f = 1.5$
End of longitudinal fillet weld, $K_f = 2.7$

TABLE AT 30   DESIGN STRESSES FOR WELDED JOINTS

(a) Recommendations by Jennings.[19.1] (b) AWS Building Code. (c) AWS Bridge Code. (d) Lincoln Electric.[19.9] (e) Adapted from AWS Bridge Code and U.S. Steel data by Blodgett;[19.16] for a common class of structural steel or the equivalent; for $n_c = 2 \times 10^6$ cycles and $n_c = 10^5$ cycles; $R = $ stress ratio $= s_{min}/s_{max}$, § 4.5. Maximum values for structural steel, butt welds, beads left on, say 18 ksi; for alloy steel, 54 ksi; for fillet welds, 12.5 and 37 ksi, respectively. The source does not mention stress concentration factors. (f) ASME Code, class 3, double-weld butt. (g) *Ultimate* strength of aluminum alloy welded with 1100 wire, argon shielded.[19.15] Use factor of safety. (h) AISC Building Code,[5.34] with E60XX electrodes.

| SHIELDED WELDING | | UNSHIELDED WELDING | | FATIGUE DESIGN STRESSES(e) | | | |
| Steady | Reversed | Steady | Reversed | Struct. Steel | | Q&T Alloy Steel | |
				2×10^6	10^5	2×10^6	10^5
16(a) 11(g)	8(a)	13(a)	5(a)	$\dfrac{16}{1-0.8R}$	$\dfrac{18}{1-0.5R}$	$\dfrac{16.5}{1-0.8R}$	$\dfrac{31}{1-0.6R}$
18(a)	8(a)	15(a)	5(a)	$\dfrac{18}{1-R}$	$\dfrac{18}{1-0.5R}$		
10(a) 15(d)	5(a)	8(a)	3(a)	$\dfrac{9}{1-0.5R}$	$\dfrac{13}{1-0.5R}$		
14(a) 11.3(g)	5(a)	11.3(a)	3(a)				
16(d) 12.7(g)				$\dfrac{7.2}{1-0.5R}$	$\dfrac{12.5}{1-0.5R}$	$\dfrac{9}{1-0.8R}$	$\dfrac{20.5}{1-0.6R}$

EQUATIONS FROM THE TEXT

(1.3) $$s = \frac{E\delta}{L} \text{ psi} \quad \text{or} \quad \frac{F}{A} = \frac{E\delta}{L} \quad \text{or} \quad \delta = \frac{FL}{AE}$$

(1.13) $$\theta = \frac{TL}{JG} \text{ radians}$$

(1.15) $$\text{hp} = \frac{(F)(2\pi r)(n)}{(12)(33,000)} = \frac{Frn}{63,000} = \frac{Tn}{63,000} = \frac{Fv_m}{33,000} = \frac{Fv_s}{550}$$

(4.3) $$q = \frac{K_f - 1}{K_t - 1} \quad \text{or} \quad K_f = 1 + q(K_t - 1)$$

(4.4) $$\frac{1}{N} = \frac{s_m}{s_y} + \frac{K_f s_a}{s_n} \qquad s_e = \frac{s_y}{N} = s_m + \left(\frac{s_y}{s_n}\right) K_f s_a$$

(4.8) $$U = \frac{F\delta}{2} = \frac{k\delta^2}{2} = \frac{T\theta}{2} \text{ in.-lb. or ft-lb.}$$

(5.1) $$F_e = \frac{s_y}{6}(A_s)^{1/2}A_s = \frac{s_y A_s^{3/2}}{6} \qquad\qquad [D < \tfrac{3}{4} \text{ in.}]$$

(5.2) $$T = CDF_i \text{ in.-lb.} \quad \begin{cases} \text{As received}^{[5.10]} & C = 0.20 \\ \text{Lubricated}^{[5.14]} & C = 0.15 \end{cases}$$

(5.3) $$F_i = QF_e\left(\frac{k_c}{k_b + k_c}\right) \text{lb.}$$

(5.4) $$F_t = F_i + \Delta F_b = F_i + \left(\frac{k_b}{k_b + k_c}\right) F_e$$

(6.1) $$s_s = K\frac{16T}{\pi D_w^3} = K\frac{8FD_m}{\pi D_w^3}$$

(6.2) $$\delta = \frac{\theta D_m}{2} = \frac{8FD_m^3 N_c}{GD_w^4} = \frac{8FC^3 N_c}{GD_w} \text{ in.}$$

(6.3) $$\frac{1}{N} = \frac{s_{ms}}{s_{ys}} + \frac{s_{as}}{s_{no}}\left(2 - \frac{s_{no}}{s_{ys}}\right) = \frac{s_{ms} - s_{as}}{s_{ys}} + \frac{2s_{as}}{s_{no}}$$

(7.1) $$F = \frac{\pi^2 EA}{N(L_e/k)^2} = \frac{\pi^2 EI}{NL_e^2} \quad \text{or} \quad s_e = \frac{F}{A}\left[\frac{s_y(L_e/k)^2}{\pi^2 E}\right] = \alpha \frac{F}{A}$$

(7.2) $$F_c = s_y A\left[1 - \frac{s_y(L_e/k)^2}{4\pi^2 E}\right] \quad \text{or} \quad \frac{F}{A} = s_e\left[1 - \frac{s_y(L_e/k)^2}{4\pi^2 E}\right]$$

(8.1A) $$\sigma = s_1 \pm s_2 = \frac{F}{A} \pm \frac{Mc}{I} = \frac{F}{A} \pm \frac{Fec}{I}$$

(8.2) $$\sigma = \frac{s}{2} \pm \left[s_s^2 + \left(\frac{s}{2}\right)^2\right]^{1/2}$$

(8.3) $$\tau = \tfrac{1}{2}(\sigma_{\max} - \sigma_{\min})$$

(8.4) $$\tau = \left[s_s^2 + \left(\frac{s}{2}\right)^2\right]^{1/2}$$

(8.10, 8.11) $$\frac{1}{N} = \left[\left(\frac{s}{s_y}\right)^2 + \left(\frac{s_s}{s_{ys}}\right)^2\right]^{1/2} \qquad\qquad \frac{1}{N} = \left[\left(\frac{s}{s_n}\right)^2 + \left(\frac{s_s}{s_{ns}}\right)^2\right]^{1/2}$$

(w) $$s_e = \frac{s_n}{s_y} s_m + K_f s_a \qquad\qquad \text{[on Fatigue Basis]}$$

(e) $$T_f = Fr = \frac{\mu Av}{h} r = \frac{\mu\pi DLv_{\text{ips}}}{c_d/2} r = \frac{4\mu\pi^2 r^3 L n_s}{c_r} \text{ in.-lb.}$$

(11.4) $$Q = h_{cr} A_b \Delta t_b \text{ ft-lb/min.}$$

$$\mathbf{h}_c = 0.017 \frac{v_a^{0.6}}{D^{0.4}} \qquad\qquad \mathbf{h}_r = 0.108 \text{ ft-lb/min-sq. in.-°F}$$

$$(12.1) \qquad \left(\frac{F_1}{F_2}\right)^k = \frac{F_1{}^k}{F_2{}^k} = \frac{B_2}{B_1} \quad \text{or} \quad \frac{F_1}{F_2} = \left(\frac{B_2}{B_1}\right)^{1/k}$$

$$(12.3) \qquad \frac{B}{B_{10}} = \left[\frac{\ln (1/P)}{\ln (1/P_{10})}\right]^{1/b} \quad [\ln (1/P_{10}) = \ln (1/0.9) = 0.1053]$$

$$(13.6) \qquad F_s = \frac{sbY}{K_f P_d}$$

$$(\mathbf{k}) \qquad F_d = \frac{600 + v_m}{600} F_t \text{ lb.} \qquad \begin{cases} \text{Commercially cut} \\ v_m \gtreqless 2000 \text{ fpm} \end{cases}$$

$$(\mathbf{l}) \qquad F_d = \frac{1200 + v_m}{1200} F_t \text{ lb.} \qquad \begin{cases} \text{Carefully cut} \\ 1000 < v_m < 4000 \text{ fpm} \end{cases}$$

$$(\mathbf{m}) \quad \text{[AGMA]} \qquad F_d = \frac{78 + v_m{}^{1/2}}{78} F_t \text{ lb.} \qquad \begin{cases} \text{Precision cut} \\ v_m > 4000 \text{ fpm} \end{cases}$$

$$(\mathbf{n}) \quad \text{[AGMA]} \qquad F_d = \frac{50 + v_m{}^{1/2}}{50} F_t \text{ lb.} \qquad \begin{cases} \text{Commercially} \\ \text{Hobbed or Shaved} \end{cases}$$

$$(13.7) \qquad F_d = F_t + \frac{0.05 v_m (bC + F_t)}{0.05 v_m + (bC + F_t)^{1/2}} \text{ lb.}$$

$$(13.8) \qquad F_w = D_p b Q K_g,$$

$$(\mathbf{r}) \qquad Q = \frac{2D_g}{D_g + D_p} = \frac{2N_g}{N_g + N_p} = \frac{2m_g}{m_g + 1}$$

$$(14.1) \qquad F_d = F_t + \frac{0.05 v_m (F_t + Cb \cos^2 \psi) \cos \psi}{0.05 v_m + (F_t + Cb \cos^2 \psi)^{1/2}} \text{ lb.}$$

$$(14.3) \qquad F_w = \frac{b D_p Q K_g}{\cos^2 \psi} \text{ lb.}$$

$$\text{For: } \begin{array}{llll} \phi_n = 14\tfrac{1}{2}° & Y = 0.314 & \phi_n = 25° & Y = 0.470 \\ \phi_n = 20° & Y = 0.392 & \phi_n = 30° & Y = 0.550 \end{array} \quad \begin{cases} \text{Worm} \\ \text{Gears} \end{cases}$$

$$(16.2) \qquad F_w = D_g b K_w$$

$$(16.3) \qquad e = \tan \lambda \left[\frac{\cos \phi_n \cos \lambda - f \sin \lambda}{\cos \phi_n \sin \lambda + f \cos \lambda}\right] = \tan \lambda \left[\frac{\cos \phi_n - f \tan \lambda}{\cos \phi_n \tan \lambda + f}\right]$$

$$(16.4) \qquad f = \frac{0.155}{v_r{}^{0.2}} \quad \text{or} \quad f = \frac{0.32}{v_r{}^{0.36}} \qquad \begin{bmatrix} \text{Bronze} \\ \text{Gear} \end{bmatrix}$$
$$[3 < v_r < 70 \text{ fpm}] \qquad [70 < v_r < 3000 \text{ fpm}]$$

$$(17.2) \qquad F_1 - F_2 = bt \left(s - \frac{12\rho v_s{}^2}{32.2}\right)\left(\frac{e^{f\theta} - 1}{e^{f\theta}}\right)$$

$$(17.3) \qquad L \approx 2C + 1.57(D_2 + D_1) + \frac{(D_2 - D_1)^2}{4C}$$

$$(17.4) \qquad \theta = \pi \pm 2 \sin^{-1} \frac{R - r}{C} \approx \pi \pm \frac{D_2 - D_1}{C} \text{ radians}$$

$$(\mathbf{1}) \qquad C = \frac{4r \sin \theta/2}{\theta + \sin \theta} = \frac{2D \sin \theta/2}{\theta + \sin \theta}$$

$$(18.2) \qquad T_f = \frac{fN(r_o + r_i)}{2} = fN r_m$$

$$(\mathbf{g}) \qquad Fe = \frac{s_{s1} J_C}{\rho'} \quad \text{or} \quad s_{s1} = \frac{Fe\rho'}{J_C}$$

$$(\mathbf{h}) \qquad J_{C1} = \bar{J} + Ad^2 = \frac{AL^2}{12} + Ar^2$$

SOME USEFUL EQUATIONS OF ANALYTIC MECHANICS

It is assumed that the reader will recall the correct manner of using these equations.

A. RESULTANT OF COPLANAR FORCES

$$R = [(\Sigma F_x)^2 + (\Sigma F_y)^2]^{1/2}$$

at an angle with the x axis of

$$\theta = \tan^{-1} \frac{\Sigma F_y}{\Sigma F_x}$$

For nonconcurrent forces, the location of the line of action of the resultant is found from the principle that the sum of the moments of any number of forces with respect to any convenient axis is equal to the moment of their resultant with respect to the same axis. If $R = 0$, but $\Sigma M \neq 0$, the resultant is a couple whose magnitude is ΣM_o, where O is any convenient center of moments.

B. CONDITIONS OF EQUILIBRIUM, COPLANAR FORCES

$$\Sigma F_x = 0 \qquad\qquad \Sigma F_y = 0 \qquad\qquad \Sigma M = 0$$

C. CENTROIDS AND MOMENTS OF INERTIA

$$A\bar{x} = \Sigma A'x' = \int x\, dA \qquad A\bar{y} = \Sigma A'y' = \int y\, dA$$
$$m\bar{x} = \Sigma m'x' = \int x\, dm \qquad m\bar{y} = \Sigma m'y' = \int y\, dm$$
$$m\bar{z} = \Sigma m'z' = \int z\, dm$$

$$AREAS \qquad\qquad MASSES$$

$$I_x = \int y^2\, dA \qquad\qquad I = \int r^2\, dm$$
$$I = \bar{I} + Ad^2 \qquad\qquad I = \bar{I} + md^2$$
$$k = \left(\frac{I}{A}\right)^{1/2} \qquad\qquad k = \left(\frac{I}{m}\right)^{1/2}$$

D. KINEMATICS

$$v = \frac{ds}{dt} \qquad\qquad \omega = \frac{d\theta}{dt}$$

$$a = \frac{dv}{dt} = \frac{d^2s}{dt^2} = \frac{v\, dv}{ds} \qquad \alpha = \frac{d\omega}{dt} = \frac{d^2\theta}{dt^2} = \frac{\omega\, d\omega}{d\theta}$$

$$a_n = \frac{v^2}{r} = r\omega^2 = v\omega \qquad a_t = r\alpha$$

E. FORCE AND INERTIA RELATIONS

Body in Rectilinear Translation:

$$\Sigma F_x = m\bar{a}_x \qquad \Sigma F_y = m\bar{a}_y \qquad \Sigma M_o = 0$$
$$[\text{INERTIA FORCES NOT INCLUDED IN THE FREE BODY}]$$

where the point O is (1) the mass center, or (2) on the line of action of the resultant force.

152

Body in Rotation and Plane Motion:

$$\Sigma F_x = m\bar{a}_x \qquad \Sigma F_y = m\bar{a}_y \qquad \Sigma M_o = I_o\alpha$$
[INERTIA FORCES NOT INCLUDED IN THE FREE BODY]

where the point O is (1) a fixed center of rotation, (2) the mass center, (3) a point whose total acceleration is directed through the mass center, or (4) a point whose acceleration is zero.

F. WORK AND ENERGY

$$U = \int F\, ds$$

For rectilinear translation:

$$U_{net} = \int R\, ds = \Delta KE$$

$$KE = \frac{Wv^2}{2g_o}$$

For rotation:

$$U_{net} = \int M_o\, d\theta = \Delta KE$$

$$KE = \frac{I_o\omega^2}{2}$$

For plane motion:

$$KE = \frac{W\bar{v}^2}{2g_o} + \frac{\bar{I}\omega^2}{2}$$

where $\bar{I}$ is with respect to the mass center; or

$$KE = \frac{I_o\omega^2}{2}$$

where the point O is the instantaneous center of rotation.

G. IMPULSE AND MOMENTUM

$$\int R_x\, dt = \Delta m\bar{v}_x \qquad \int R_y\, dt = \Delta m\bar{v}_y$$
$$\int M_o\, dt = \Delta I_o\omega$$

where the point O is (1) the instantaneous center of rotation, (2) the mass center, or (3) a point whose velocity is directed through the mass center.

SOME USEFUL MATHEMATICAL RELATIONS

NOTE. No attempt has been made to include here all mathematical relationships that may be needed in the solution of problems in this book. It is presumed that the reader will recall the correct manner of using these equations.

TRIGONOMETRY

Law of sines: $\dfrac{\sin A}{a} = \dfrac{\sin B}{b} = \dfrac{\sin C}{c}$

Law of cosines: $a^2 = b^2 + c^2 - 2bc \cos A$

Relations between rectangular and polar coordinates:

$$x = \rho \cos \theta \qquad y = \rho \sin \theta \qquad \rho = (x^2 + y^2)^{1/2} \qquad \theta = \arctan \frac{y}{x}$$

Functions of angles:

$$\sin 2\theta = 2 \sin \theta \cos \theta \qquad \sin^2 \theta + \cos^2 \theta = 1$$
$$\cos 2\theta = \cos^2 \theta - \sin^2 \theta = 2 \cos^2 \theta - 1 = 1 - 2 \sin^2 \theta$$

$$\sin\left(\frac{\theta}{2}\right) = \left(\frac{1 - \cos \theta}{2}\right)^{1/2} \qquad \cos\left(\frac{\theta}{2}\right) = \left(\frac{1 + \cos \theta}{2}\right)^{1/2}$$

$$\sin^2 \theta = \frac{1 - \cos 2\theta}{2} \qquad \cos^2 \theta = \frac{1 + \cos 2\theta}{2}$$

$$\sin (\alpha \pm \beta) = \sin \alpha \cos \beta \pm \cos \alpha \sin \beta \qquad \sin (90° \pm \alpha) = \cos \alpha$$
$$\cos (\alpha \pm \beta) = \cos \alpha \cos \beta \mp \sin \alpha \sin \beta \qquad \cos (90° \pm \alpha) = \mp \sin \alpha$$

$$\tan (\alpha \pm \beta) = \frac{\tan \alpha \pm \tan \beta}{1 \mp \tan \alpha \tan \beta} \qquad \tan (90° \pm \alpha) = \mp \cot \alpha$$

Functions of frequently used angles:

	0°	14½°	15°	20°	30°	45°	60°	75°	90°
sin	0	0.250	0.259	0.342	0.5	0.707	0.866	0.966	1
cos	1	0.968	0.966	0.940	0.866	0.707	0.5	0.259	0
tan	0	0.259	0.268	0.364	0.577	1		1.732	3.73	∞

SOLUTION OF QUADRATIC EQUATION

$$Ax^2 + Bx + C = 0 \qquad x = \frac{-B \pm \sqrt{B^2 - 4AC}}{2A}$$

CALCULUS

Differentials:

$$d(xy) = y \, dx + x \, dy \qquad d\left(\frac{x}{y}\right) = \frac{y \, dx - x \, dy}{y^2}$$

$$d(\sin x) = \cos x \, dx \qquad d(\cos x) = - \sin x \, dx$$

$$d(\sin^2 x) = 2 \sin x \cos x \, dx \qquad d(\cos^2 x) = -2 \cos x \sin x \, dx$$

$$dx^n = nx^{n-1} \, dx \qquad d(\tan x) = \sec^2 x \, dx = \frac{dx}{\cos^2 x}$$

Integrals:

$$\int x^n \, dx = \frac{x^{n+1}}{n+1} + C \qquad [n \neq -1]$$

$$\int \frac{dx}{x} = \log_e x + C = \ln x + C$$

$$\int (a^2x^2 + b^2)^{1/2} \, dx = \frac{x}{2}(a^2x^2 + b^2)^{1/2} + \frac{b^2}{2a}\log_e[xa + (a^2x^2 + b^2)^{1/2}]$$

$$\int (b^2 - a^2x^2)^{1/2} \, dx = \frac{x}{2}(b^2 - a^2x^2) + \frac{b}{2a}\sin^{-1}\left(\frac{ax}{b}\right)$$

$$\int \sin\theta \, d\theta = -\cos\theta + C \qquad\qquad \int \sin^2\theta \, d\theta = \frac{\theta}{2} - \frac{\sin 2\theta}{4} + C$$

$$\int \cos\theta \, d\theta = \sin\theta + C \qquad\qquad \int \cos^2\theta \, d\theta = \frac{\theta}{2} + \frac{\sin 2\theta}{4} + C$$

$$\int e^x \, dx = e^x + C \qquad\qquad \int \sin\theta\cos\theta \, d\theta = \frac{\sin^2\theta}{2} + C$$

AREAS

Triangle, $bh/2$	Circle, πr^2 or $\pi D^2/4$
Trapezoid, $h(a+b)/2$	Ellipse, πab

VOLUMES

Parallelepiped, abh	Cone, $\pi r^2 h/3$
Sphere, $4\pi r^3/3$ or $\pi D^3/6$	Ellipsoid, $4\pi ab^2/3$

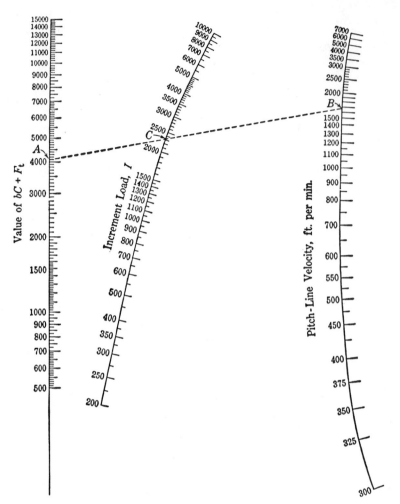

Courtesy J. G. H. Thompson

FIGURE AF 22. *Alignment Chart for Solving Buckingham's Dynamic-Load Equation.*
To use: Suppose $bC + F_t = 4000$, $F_t = 493$, and $v_m = 1675$ ft./min. Spot 4090 on the left-hand scale at A; spot 1675 on the right-hand scale at B; read $I = 2310$ on straight line AB and middle scale at C. Then $F_d = I + F_t = 2310 + 493 = 2803$ lb., approximately.

Decimal Equivalents of an Inch

Fraction	Decimal	Fraction	Decimal
$\frac{1}{64}$	0.015625	$\frac{33}{64}$	0.515625
$\frac{1}{32}$	0.03125	$\frac{17}{32}$	0.53125
$\frac{3}{64}$	0.046875	$\frac{35}{64}$	0.546875
$\frac{1}{16}$	0.0625	$\frac{9}{16}$	0.5625
$\frac{5}{64}$	0.078125	$\frac{37}{64}$	0.578125
$\frac{3}{32}$	0.09375	$\frac{19}{32}$	0.59375
$\frac{7}{64}$	0.109375	$\frac{39}{64}$	0.609375
$\frac{1}{8}$	0.125	$\frac{5}{8}$	0.625
$\frac{9}{64}$	0.140625	$\frac{41}{64}$	0.640625
$\frac{5}{32}$	0.15625	$\frac{21}{32}$	0.65625
$\frac{11}{64}$	0.171875	$\frac{43}{64}$	0.671875
$\frac{3}{16}$	0.1875	$\frac{11}{16}$	0.6875
$\frac{13}{64}$	0.203125	$\frac{45}{64}$	0.703125
$\frac{7}{32}$	0.21875	$\frac{23}{32}$	0.71875
$\frac{15}{64}$	0.234375	$\frac{47}{64}$	0.734375
$\frac{1}{4}$	0.25	$\frac{3}{4}$	0.75
$\frac{17}{64}$	0.265625	$\frac{49}{64}$	0.765625
$\frac{9}{32}$	0.28125	$\frac{25}{32}$	0.78125
$\frac{19}{64}$	0.296875	$\frac{51}{64}$	0.796875
$\frac{5}{16}$	0.3125	$\frac{13}{16}$	0.8125
$\frac{21}{64}$	0.328125	$\frac{53}{64}$	0.828125
$\frac{11}{32}$	0.34375	$\frac{27}{32}$	0.84375
$\frac{23}{64}$	0.359375	$\frac{55}{64}$	0.859375
$\frac{3}{8}$	0.375	$\frac{7}{8}$	0.875
$\frac{25}{64}$	0.390625	$\frac{57}{64}$	0.890625
$\frac{13}{32}$	0.40625	$\frac{29}{32}$	0.90625
$\frac{27}{64}$	0.421875	$\frac{59}{64}$	0.921875
$\frac{7}{16}$	0.4375	$\frac{15}{16}$	0.9375
$\frac{29}{64}$	0.453125	$\frac{61}{64}$	0.953125
$\frac{15}{32}$	0.46875	$\frac{31}{32}$	0.96875
$\frac{31}{64}$	0.484375	$\frac{63}{64}$	0.984375
$\frac{1}{2}$	0.5	1	1.0